Contents

Introduction

Measuring matters. Why?

What's good about measuring? First and foremost, although measuring is not science, every kind of science, and every kind of technology, depends on it.

From atomic structure to the invisible stars, measurement is the basis on which scientific knowledge and thinking rest.

Galileo instructed his followers some 350 years ago to 'count what is countable, measure what is measurable'—and modern advances depend on exactly the same principles though to a far greater extent.

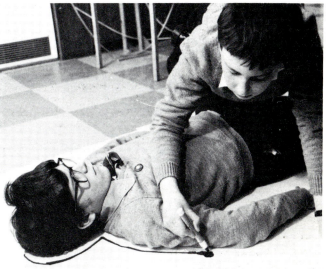

But measuring can be the most boring activity, or it can be made stimulating and even exciting. We need to know why it is worth while measuring something, we need the right tool or instrument to measure it with, and we need the chance to do it ourselves if at all possible rather than to look at diagrams or watch somebody else. Perhaps we also need to learn how to get it just right.

Measurement can and must be relevant to children's or students' current interests, or to adult interests which they share. The latter type of interest is often strong, since it matches children's perception of their own interests in the imaginable future; for example in adult height, or in consulting the dials on the dashboard of a car. Finding out measurements in real life, or looking them up if the objects are genuinely not available, can become an absorbing activity—from the changing weight of a baby to that of a football before and after a wet match, from the height of a pet pony to that of Mount Everest.

Measuring is among the vital learning activities.

This is a practical book about measuring, not about calculating. Here and there bits of simple arithmetic come in because they are needed, and there are some 'looking up' activities when actual measuring is impossible, but most of the activities are real things to do.

The chapters often begin with very easy measurements and ideas. Some of these are worth doing even if you know all about it; there may just be a new hint, or it may be explained in a new way. The later part of each chapter may not be quite so easy, but it goes a step at a time, and none of it is very difficult.

Many of the measurements are important in ordinary everyday life; they include the sizes of clothes and the measurement of time—but the

scientific background is there, and the more one does, the more competent one becomes. The skills involved, and the understanding which comes with using them, are invaluable in technology and in all the sciences.

APU test reports

The project in Science at age 11 (1980/81) from the Assessment of Performance Unit began with a list of 'testable' areas in science—not in any way to be regarded as a syllabus. Under 'Science—a mode of thought and activity', the testing programme named six specific categories, of which No. 2 is using apparatus and measuring instruments. Their list of 'measures' includes time, mass, temperature, length, area, volume and capacity, also involved in the testing programme in mathematics.

This, incidentally, may be one of the reasons for the somewhat disappointing results—mathematics teachers may think practical measuring activities are 'science', while the science teachers suppose 'measuring' to have been part of early mathematics. . .

Scientific method was also looked for, but report No. 1 says that very few pupils repeated observations or measurements as routine. 'A small minority made notes other than recording measurements.'

As many pupils of 11 will have had no specific training in the methods of work in science, this is hardly surprising.

The results of the first testing programme on 15-year-olds, reported in 1982/83, include as part of school science activities a number of short practical tasks such as the use of particular measuring instruments, reading scales, measuring small quantities accurately, and making a number of

4

14

8.90

This book is to be returned on
or before the date stamped below

Basi ities

Also in the *Hulton Practical Teaching Series*

AIR AND WATER ACTIVITIES
Dorothy Diamond with Mary Brimacombe
ISBN 0 7175 1187 1

PRE-COMPUTER ACTIVITIES
Dorothy Diamond
Teacher's Book
ISBN 0 7175 1154 5

GET READY FOR THE COMPUTER
Dorothy Diamond
Pupil's Books
Book 1   ISBN 0 7175 1335 1
Book 2   ISBN 0 7175 1336 X
Book 3   ISBN 0 7175 1337 8

First published in Great Britain 1985
by Hulton Educational Publications Ltd
Raans Road, Amersham, Bucks HP6 6JJ

Text © Dorothy Diamond 1985
Illustrations © Hulton Educational Publications Ltd
1985

ISBN 0 7175 1359 9 ✓

Set in 11/12 Linotron Times by Input Typesetting Ltd,
London
Printed in Great Britain by
The Pitman Press, Bath.

Acknowledgements

The author and publishers would like to thank the
following for their help in providing material for the
book.

Photographs:
Doug Kincaid and Peter S. Coles, p. 3
Times Newspapers Ltd., p. 61
Sunflowers, p. 62, supplied by the author

Information and References:
The Austin Morris Group
Osmiroid Educational

Cover:
Frank Watkins

Artwork:
John Hopkins

Edited and designed by Elizabeth Paren and Elizabeth
Smith

estimates of volume, area, time, length and temperature. The Unit was obviously disappointed with some of the results; the report states that there were deficiencies in the area of practical skills.

Low performance levels were found in the use of certain measurement procedures; in particular, measurement of short distances and short time periods gave mean scores below 20%, indicating that few pupils employ an appropriate technique.

Only about one-third of the pupils could read more complex types of scale. Most pupils could give reasonable estimates for lengths and for time, but they were less accurate in estimating areas, volumes, temperatures and forces.

The APU testers suggest that in spite of experience in practical science, knowledge of some of the basic skills might have been expected to be greater and more widespread. They conclude that perhaps the teaching of specific measurement skills needs to be given more attention.

The APU 1980/81 report on mathematics performance of 15-year-olds suggests something deeper; the report says that more than 60% (but this is fewer than 7 out of 10, after all) could use a ruler, a protractor and a balance for basic measuring—but that concepts relating to these measurements were not well understood.

This book attempts by means of very many different small activities to help pupils at all stages in their school career to want to measure, to do it correctly and accurately, and to understand the methods and the importance of the skills they use.

With material as varied as babies' bottles and Big Ben, sunflowers and bicycle wheels, it seems likely that most people will find something to interest them in science measuring activities. One can then say that in this case even the sky is not the limit.

S.I. Units

Since 1960, scientists have been using a special set of units for measuring and writing about measurements. This set of measurements is called the S.I. — Système International or Standard International — set of Units, and, as its name indicates, it has been agreed internationally.

The SI units used for the basic measurements in this book would be:

LENGTH: *metre* AREA: *square metre* VOLUME: *cubic metre* MASS: *kilogram* FORCE: (e.g., weight) *newton*. TIME: *second* ANGLE: *radian*

Some of these are much too big for everyday purposes, one is too small, and one is too difficult to measure with ordinary instruments. We therefore use decimal fractions or multiples of the first four, and other practical units for the last three.

Volume

Volumes of liquids and 'pourables'

Volume is the first dimension we meet in life. A cupful of milk predates a step by months, while scooping and tipping water is a very early bath-time activity.

Some of the early physical skills, though not directly scientific, are essential for later efficiency and accuracy.

Pouring without losing any liquid

Care needs to be taken
(a) not to start with too full a container;
(b) to use the lip, spout, or corner of cuboid;
(c) to have a steady action and steady hand(s);
(d) to remember/discover to let air in as well as liquid out through a bottle-neck—and at the same time;
(e) to use a funnel if necessary;
(f) to have an assistant to hold the second container and funnel upright if necessary;
(g) to watch what is happening, especially that the liquid already in the funnel is not about to make the receiving container overflow;
(h) to stop without dribbles down the side.

Some people will find this list 'a matter of common sense', but many adults could have profited from some practice in these skills earlier.

Volume and capacity

The two measurements are different, but occur together very often.

Volume—how much of it is there?
Capacity—how much will it hold?
And the linking question—will it hold it? ('Go or not-go' as in the use of engineering gauges.)

Estimation rather than exact measurement is frequently what is needed. Will a pint of milk go into a particular jug? Will a litre of solution go into a particular bottle?

Estimation can only be worth anything if it is based on experience, and the more experience(s) the better. Guesses without knowledge behind them are useless, especially in view of the effect of the shape of the container on the human mind.

Tall narrow bottles, jars and packets are used by manufacturers for their products in the confident and well-founded belief that the buyers will assume that a tall container holds more than a short (even though fat) one. A maker of salad dressing tried it as an experiment—the dumpy jars, with wide enough necks for a spoon too, simply did not sell, while the tall narrow ones (same volume inside) sold as always.

Establishing a standard volume

Activities are to be planned to establish a standard volume, to be held firmly in mind for comparisons.

The litre is the basic metric measurement for liquids, therefore this is the obvious choice for a standard volume. One litre is sometimes contracted to '1l', as on wine bottles, but this contraction is not to be encouraged since it is far too easily misread; 'litre(s)' should be written out in full.

(a) Collect plastic drinks bottles which contained 1 litre. Find as many different varieties as possible, but all should be 1 litre size.

(b) Fill each with water, preferably coloured with different food-colourings, to the original level in the neck of the bottle.

(c) Add to the collection clear plastic drinks bottles which held 1½ litres, and 2 litres. Pour 1 litre into each (from a spare 1 litre bottle), again colouring the water.

(d) Add larger transparent or translucent plastic containers, including some large bottle-type containers with hollow handles, some rectangular (picnic water-containers, etc.) but each containing 1 litre.

(e) Finally collect sets of smaller clear plastic beakers and jars, e.g. from bubble-bath liquid in numbers sufficient to contain 1 litre in each set of identical containers. Colour each litre with a separate colour, and group them in litres.

This collection makes a good display, and the experience of preparing it is most helpful in establishing the concept of 'a volume of 1 litre'.

Using standard equipment

Following on from this everyday container collection, the use of the standard set from the suppliers' catalogues is essential; i.e. a cube, a flat cuboid, a tall cylinder and a flat cylinder, each holding 1 litre (with a small extension above the litre mark to avoid spilling).

The basic item is the cube, measuring 10 cm along each edge (11 cm upwards), since 1 litre is the volume of liquid in 1 000 cubic centimetres. Mathematics apparatus often includes a 10 cm cube, which can be packed with 1 000 centimetre cubes for demonstration. This is an exact volume; in the case of the containers for liquids, it is a good idea to mark the 1-litre (1 000 millilitres) level with a coloured sticky label.

litre containers

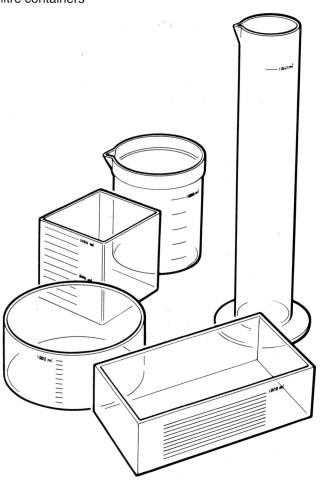

These containers are useful for demonstrating that 1 000 cubic centimetres and 1 000 millilitres are the same volume—the cube can be seen to contain the one, and is marked with the other.

7

For adults, even highly intellectual ones, and for children, simply to pour 1 litre of (preferably coloured) water from one standard container to another is a genuinely instructive experience if they have not had it before. It *is* difficult to believe that the 1 000 millilitres in the tall measuring jar (cylinder) will really go into the flat bowl or the 10 cm cube. But after several personal trials—to and fro—8-year-olds become confident that it *will* go in, even though it still doesn't look as if it will.

The concept of *conservation of volume*—the comprehension that a certain, given, amount of liquid stays the same 'size' no matter what shape the container, is slow to develop—for some it never arrives—and cannot be hurried by 'teaching'. However, its achievement can be greatly helped by practical activities, spread out over several years.

Using 'pourable' solids

The containers which hold 1 litre of water can equally well be used to hold other 'pourables', for example, fine dry sand, salt, rice, lentils, dried peas . . . so that the concept of volume is not attached only to water. In the case of pourable solids, of course, the 1 000 cubic centimetres version is used.

Finding the volume of gases

This tends to involve the use of a closed container, which will be filled by the gas, or the gas being collected 'over water', with the opening at the bottom of the container. Here the usual markings on the jar or bottle will be inverted, but they only need to be read downwards from 0 cu cm/ml rather than upwards.

The temperature makes much more difference to the volume of a gas than to the volume of a liquid, so that air, say, collected over warm water will (until it cools down) have a larger volume than it would when cold. In the same way, pressure makes a considerable difference to the volume of a gas, while water is almost incompressible under room conditions. The jam jar with a balloon rubber sheet and a pointer on the open top, sometimes proposed as a classroom model barometer, is likely to be more influenced by room temperature than by atmospheric pressure.

Accuracy in measuring volumes of liquids

Accuracy can be very important indeed, but it depends on only a few factors. These can be investigated practically, by simple activities.

(a) *Getting the measuring container upright.* Half-fill a transparent plastic drinks bottle—1 litre or larger—with coloured water. Stand it carefully on a horizontal table, and slide a rubber band down on the jar to the level of the top of the water (the water level). Check by getting down until eye-level is at water level. This is the only accurate way to look at it. Now tilt the bottle a little one way (look at the water level and the rubber band marker), then further in the other direction. The container must be vertical for useful level—therefore volume—observations.

notice the meniscus

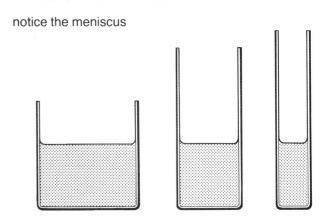

8

(b) *Measure from the centre of the water surface.* Using the same plastic bottle, look horizontally across the water surface. There are two apparent water levels; one (the higher) is where the water curves up all round the edge of its surface, clinging to the inside of the bottle. The second level, a little lower, goes right across the water, absolutely flat, until just before it curves up again on the far side. This is the level to be measured, since the curved edge only takes up a tiny fraction of the total volume. With some kinds of plastic, water hardly clings at all, and therefore hardly rises at the edges.

The curved surface is the meniscus. In a narrow tube the whole surface may be curved, with no space for a flat centre. In this case the lowest level, in the centre of the curve, is taken for measurement.

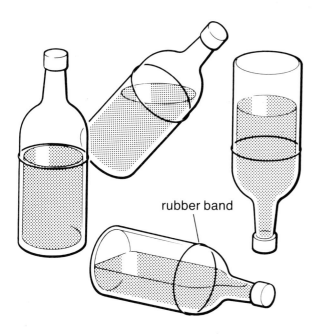

rubber band

(c) *Use a narrow container rather than a wide one.* A few drops more will make a visible difference in a narrow tube such as a measuring jar (cylinder) of the laboratory type, but be unnoticed in a wide container. Try this by marking the water level in a narrow and a wide container, each about half-full, either with a thin rubber band round each, or a wax pencil or an old lipstick. . . Get the marks as exact as possible; then add to each container equal small volumes of water—say 10 drops from a dropper. Look for the difference which this addition has made to the level in the narrow and the wide container.

The greater accuracy gained by using a narrow measuring jar is the reason why these tall jars are always used in science, in spite of the fact that many of them get tipped over or knocked and broken accidentally.

(d) *Get eye-level and water-level at the same height.* Mark the water level in a clear container such as a clear plastic drinks bottle, either with a thin rubber band or a wax pencil, etc. Look at the level very carefully, with eye-level and water-level at the same height. This usually involves sitting, or bending the knees; it will *not* do to pick up the container and hold it, as two kinds of inaccuracy may come in—slope and wobble. Then look down at the water-level and marker from higher up, and finally, by crouching, from below the water-level. In each position the apparent level can be seen to be inaccurate.

(e) *Take into account the actual shape of the measuring container.* Bottles, jars and measuring cylinders very often have a curved surface in the base, so that the markings up the side either do not start right at the bottom, or are put on by experiment, allowing for this curvature. Test this with one of the clear

9

plastics drinks bottles; pour in a number of equal volumes of coloured water, marking the level after each addition. The distance from the apparent bottom of the bottle to the first mark up the side will be clearly greater than the distance between two markings that represent an equal volume addition further up.

Measurement of very small volumes

How much water is there in a drop? Since it is much less than 1 millilitre, the method is to measure the volume of a fairly large number of drops and then divide the total by the number. It helps, therefore, to decide on sensible 'round numbers', e.g. 100. A small measuring jar is usually marked in millilitres, and 100 drops from a dropper will be a measurable volume. If there are no markings at the bottom of the jar, one can always put in some water up to a noted mark, then add the drops and measure how much extra has been added. This is, after all, a method recommended for weighing small children (and even dogs) on weighing machines meant for heavier people. 'Stand on the platform yourself, note your own weight, then take child (or dog) on with you, and see how much the weight has increased.'

In any case, the more drops you have the patience/ scientific determination to count, the more accurate the result will be.

With small volumes of water, one has to take into account water already in the container from a previous measurement—a wet jar—and/or the water left behind when water is poured out of a container into the measuring jar.

Are all drops the same size? Test drops of water containing detergent. Count and measure drops from different sized openings, e.g. taps, well-washed washing-up liquid bottles, plastic beakers with different sized holes made in them. How reliable are instructions to 'use five drops' of this or that? Look for examples.

Rain gauges

Rainfall is measured in a special way, i.e. as the depth of water falling on the land surface, so not as a volume. In most places in Britain the rainfall is so small that rain gauge containers are marked from 0 to 50 mm (and 2 inches), or only 0–10 mm, in 0.1 mm divisions. Such small measurements are

rain gauges

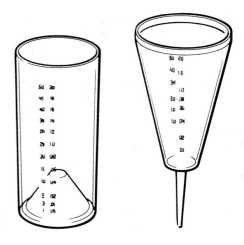

made possible by having containers with very narrow bases, either funnel-shaped or with an internal raised cone to 'spread' the water upwards.

The use of rain gauges can easily be practised without waiting for the weather by using a lawn sprinkler; this is also a method of checking the watering effect produced by the sprinkler compared with rainfall.

Dipsticks

A dipstick measures the depth of a liquid, therefore only indirectly the volume; however, the measurement of volume is the main use. Each dipstick must be marked, or calibrated, for a particular type of container, and for the information needed. It is not necessary to know the exact volume of oil in a tank, or tanker, but it is very important to know whether it is full, half-full or empty—and it may be quite impossible to see the level, or even the surface, directly.

Dipsticks are used to gauge the volumes of liquids such as oil in the car sump, petrol in the large tanks of garage forecourts, beer at the brewery, and dry-cleaning fluid at the suppliers, among other uses.

The instructions in a car manual make the method of measurement clear and accurate, e.g. (for the MG Metro): stand the car on level ground, let the engine cool, take out the dipstick and wipe it, put it into the sump as far as it will go, take it out again, and see how far up the dipstick the oil comes. The last is easy, since the oil sticks to the brass rod, which has the levels marked on it. At the moment many dipsticks in commercial use are still marked in Imperial units, but they are being converted to metric values, as are petrol pumps.

Simple dipstick measurements can be practised in 10 cm cube containers which are already marked

up the side in 100 ml volume levels. Each of these graduations therefore represents 1 cm in depth. Two sources of possible inaccuracy must be noted:

(a) The ruler used as a dipstick itself displaces some liquid up the sides and up the dipstick, giving a larger-than-correct reading for the volume of liquid. For an accurate result the volume of the dipstick in the liquid can be found.

(b) Most ordinary rulers do not begin at the zero marking, hence no accurate depth can be measured with them. Dead-length rulers exist, and are the only useful ones in this context (as in a number of other situations). Most dead-length rulers in classroom use are wooden, which makes them unsuitable for much work in water. Good plastic dead-length rulers, with markings (cm or cm and mm) horizontally on one side, and vertically (just right for dipsticks) on the other side, are made by Taskmaster.

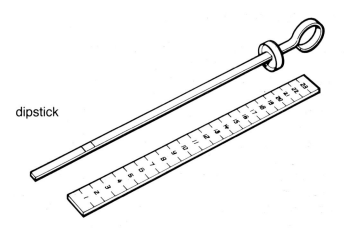

dipstick

Dipstick-type measurements are essential in field-work studies of ponds, rockpools, etc. but the depths are often greater than the length of a ruler. Two possible methods, also used in the commercial world, are to use metre or longer rods, marked in cm or perhaps 10 cm units, or to use a waterproof tape-measure weighted at the zero end. This ensures that it hangs straight down, and indicates when it reaches the bottom. It would, of course, be unsuitable to use the surveyors' roll-in tape unless it were completely waterproof, and were thoroughly dried before being run back into its case.

For accuracy and certainty, dipstick measurements should be repeated at least once; car owners are often seen to wipe the oil dipstick and put it in again—just checking. . .

Measuring or measuring out

So far we have considered measuring the volume of a given quantity of liquid. Scientific activities as well as everyday ones often need the measuring out, from a 'reserve', of a specific volume. This has to be measured by different methods, sometimes with different apparatus.

Measuring out rough (approximate) volumes. There are plenty of examples where the volume does not need to be accurately measured. In such cases it is good common sense not to spend time getting the volume to the nearest millilitre, e.g. 'Half fill a beaker with cold water.' 'Add water until the fish-tank is about three-quarters full.' 'Mix the powder with three or four drops of water.'

Measuring out precise volumes. However, it is often necessary in scientific work to measure out a volume very precisely, in which case the apparatus may be, for example, a pipette, a burette, a syringe or a specially made and marked dropper. There are two main ways to get the right volume.

(a) To fill the graduated container (measuring jar etc.) to the mark for the required volume, so that it just contains the volume needed. This can then be poured out completely.

(b) To start with any volume larger than the amount needed. Note the level in the container, pour out carefully until near the calculated lower level, stand the container on a flat surface to check, pour out a little more, check again, and continue until the level shows that exactly the right volume has gone. This second method may be the only one available, for example if the first (larger) volume cannot be controlled accurately, but it needs much more care than method (a), especially as it may well be impossible to go back if too much liquid has been poured out of the graduated container.

On a small scale, a pipette provides an example of method (a); rather more liquid is drawn up into the pipette than is required, either by mouth (if harmless) or by pipette filler. Then by moving the finger which closes the top of the pipette slightly to and fro across the top, drops of liquid are allowed to fall out into the reserve container until the level in the vertically held tube is exactly at the mark. The lowest point of the surface is, of course, observed; the curve of the liquid surface (the meniscus) in such a narrow tube is almost a semicircle in side view.

Once the contents of the pipette are exactly the required volume, the finger is taken off and all the liquid is allowed to run out where it is needed. There will be a drop or two left in the tip of the pipette; this has been allowed for when the mark was put on the tube, so do not blow it out.

A burette gives an example of method (b). The burette is fixed vertically in its stand, and is filled up to the zero marking near the top, or to any other mark well above the volume needed. This is done by putting in some excess, e.g. above the zero, and running out a little through the tap (or the clip on the rubber tubing) into the reserve vessel, so that the jet at the bottom is full before the measured

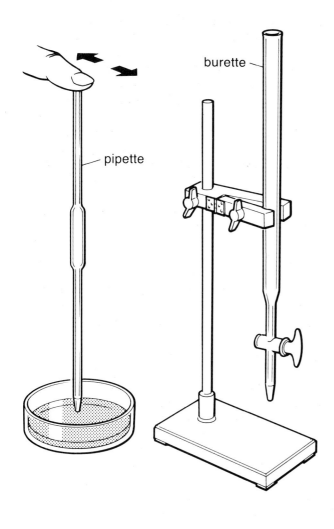

pipette

burette

minimal apparatus, e.g. the pipette management with plastic straws, and the burette drop by drop adjustment with glass or plastic droppers.

Medical syringes (without injection needles) can be used instead of pipettes to deliver exact, known, volumes of liquid; filling the tip, to displace the air, is well-known to be important.

Volumes of solids
Regular solids

The volume of a regular solid e.g. a cuboid, is usually found by measuring length, breadth and height, and multiplying. Volumes of solids are correctly expressed in cubic metres (very large) or cubic centimetres (very small). The best way to get used to them is to handle many centimetre cubes e.g. Centicubes, or similar plastic or wooden cubes without the linking projections and cavities. In the same way, a framework of a 1 metre cube can be built of linking rods or of garden canes (slightly over 1 metre long) fixed together with rubber bands at the corners. This large cube presents a challenge. Who can get inside it? How many can get inside it? How does its volume compare with that of a telephone box?

Measurement by displacement of water

For objects of other shapes (including spheres, whose volume formula should not just be assumed), the best method of measurement is by displacement of water—as long as the solid is insoluble and undamaged by water. For soluble materials some other liquid such as paraffin can be used.

1. The first point to be established is that a solid object lowered into water will only sink if it is denser than water; a special method is necessary if it floats. Begin with materials which sink!

volume is let out. If this is not done, the first liquid coming through the tap will only fill the jet, and the volume delivered will be less than the markings show. The tap or clip can be opened fully at the beginning if a fairly large volume is needed, but the level in the burette must be watched carefully and constantly, and the flow cut down to drops when the limit is near.

Simple technique skills can be practised with

2. The second—and this seems obvious, but is not, unless backed by experiences—is that when a solid object goes into water, it displaces its own volume of water upwards. Remember that Archimedes is said to have made this observation with great excitement!

(a) A hand, perhaps fist, lowered into the water in a wide-necked jar is a suitable substitute. It can be raised and lowered a number of times, and the levels of the water before, during and after the immersion can be recorded on the side of the jar for comparison. Even if the jar is not graduated, the volume of the hand can easily be found by taking the hand out and adding water from a measuring jar to fill the unmarked container up to the level reached when the hand was in.

(b) Graduated measuring jars tend to be narrow (for reasons of accuracy) so that objects for testing must be carefully chosen. At one time the textbook object was a glass stopper on a thread; unfortunately it went in, almost to the bottom, but far too often became wedged. . .

For early experiences of displacement and measurement of water volume displaced, a green Plasticine 'man' with a thread round his neck is satisfactory, and memorable. Lowering and raising 'him' slowly shows the gradual displacement of water upwards in the jar, and its later fall, back to its original level.

(c) Smaller objects, such as glass marbles, tend to have volumes which are too small for accurate measurement. As in other situations where single objects cannot be correctly measured, one takes a definite number—say 10—and then divides the result. The glass marbles must, of course, all be the same size. Care needs also to be taken over dropping them into the measuring jar, unless it is a strong plastic one.

3. The volume of an object which floats in water is found by using another object dense and heavy enough to sink it. The water level is adjusted with the heavy object already in the water; the light one is then tied to the dense one, and the two are lowered into the water together. The new level gives the volume of water displaced by the second, less dense, object.

4. Instead of noting two levels in the measuring vessel, the actual water displaced by a sunk or floating object can be found directly by using a displacement can or bucket (a can for small objects, a bucket for larger ones e.g. a brick or a football). The container is filled with water until some runs out of the side spout or lip. When no more runs out, a graduated container is placed under the spout, and the object is gently lowered into the water. Whether it floats or not, the water it displaces runs over into this container; from here the displaced water can be measured—giving the volume of a sunk object, or the 'underwater' volume of a floating object.

The displaced water could equally well be weighed, which is what would be needed to determine the upthrust on the object when partly or completely in the water. This upthrust makes the difference between its true weight (in air) and its apparent weight, which is less, when in or partly in water.

A simple, free, displacement vessel can be made from a large yoghurt pot by heating part of the top edge in steam from a kettle, and with tongs pulling the softened rim out and downwards a little. A smaller yoghurt pot below the 'lip' will catch the overflow water.

With some small displacement cans, the surface tension of the water at the outlet holds it back when it should overflow. This can be remedied by adding a drop of detergent, which makes no other difference to the result.

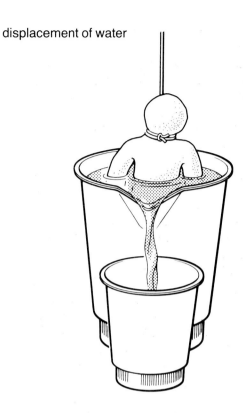

displacement of water

Volume of air

The volume of the spaces in between
How full is full?

(a) Fill a clear plastic beaker with as many glass marbles as will go in, with as nearly a flat top as possible. The beaker is 'full' of marbles, but there is a good deal of space between them. How much space? From a measuring jar filled with water to a known level, pour water into the beaker of marbles until the water comes to the top edge. Stand the measuring jar down, and find the volume of water which has been poured in between the marbles. What did this water displace? Air, of course.

(b) Repeat this test with an identical beaker,

'filling' it this time with dry sand (or fine gravel). The grains of sand are so much smaller than the marbles that one can make a prediction as to the difference in the volume of 'space' which will be filled with water. Test it. The volume of air in different soils is very important.

(c) An alternative demonstration is to begin as in (a) with marbles, then, since the beaker is clearly not 'full', to pour in dry sand on top, until it is more nearly 'full'. Finally add water, measuring the volume of water which brings the beaker's contents to a beaker-full. In each case, the volume being measured is equal to the volume of the air in the spaces; such measurement is very important, e.g. in soil science, in packaging, and in heat insulation—where the air content matters.

(d) The volume of air from a particular container or material is often collected 'over water', i.e. by letting the air rise, under water, into a measuring jar full of water and inverted, with its open end still in the water. Try finding the volume of air in what is called 'an empty drinks can'. It wasn't empty, of course; by pushing it down into water with the opening downwards, and then tilting it slightly so that the air escapes under a jar full of water with the opening placed to catch the bubbles as they rise, the air can be captured. Now the level of the water left in the measuring jar can be observed (upside down). This gives a direct reading of the volume of air from the can.

The volume of a breath

One activity in which a volume of air is measured is very frequently described in primary 'science' textbooks; it is unfortunately usually headed 'How much air is there in your lungs?' A moment's thought would show that this is impossible to test, since 'you' cannot flatten 'your' lungs like a plastic bag to squash out all the air. The experiment to find the volume of air 'you' can breathe out is a good one, though not too easy simply because the volume is large. The apparatus consists of about 1 metre of plastic (or rubber) tubing, a sink (or fish-tank) about half-full of water, and a large (at least 3-litre) clear plastic sweet-jar or drinks container. If the container is large enough, there may be a problem in turning it upside down, full to the very top, into the water in the sink, without losing any water. One breath is blown into the container, displacing water into the sink or fish-tank, which must be able to hold this extra volume. The transparency of the apparatus is important, so that the effect of blowing in one steady breath can be observed. The container is then closed again, after removing the tube, and is turned right-way up. The simplest way to find the volume of

finding the volume of one breath

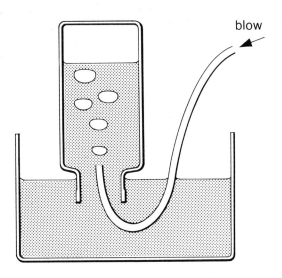

blow

the breath contained is to fill the container up again with a measuring jar, counting the volumes added. Results will vary fairly widely, from one individual to another, and even from one breath to another from the same person (who should make a break between attempts. . .).

Estimating cubic capacity

Even with practice this is often very difficult. Advertisements for fridges and freezers confidently state the cubic capacity, but as with other containers of irregular shape, it is almost impossible to measure this quantity. An enterprising car manufacturer filled a new model absolutely full of table-tennis balls to demonstrate how much space there was inside it.

Units of volume
Metric and Imperial measures

These exist side by side, often on the same container, e.g. milk bottles '1pt 568 ml'. Scientists always use the metric system, but it is important to look for metric measures in everyday life, and to get really used to them. Petrol pumps give both gallons and litres (1 gallon = 4.55 litres) and the prices for each; paint is sold in litres, wine in centilitres (a Continental habit—70 cl = 700 ml = 0.7 litre to the wine-bottle). Beer at the bar comes in pints, but in cans there are several different measures.

Much of the problem of the slow change-over is obviously due to the containers, e.g. millions of pint milk bottles. Where the numbers are smaller, the change has gone faster; for example, babies' bottles are in 120 or 240 ml sizes, though they still have 'liquid oz.' on the side. Unlike salad cream bottles, babies' bottles have good reason to be rather tall and narrow.

Practical, if unscientific, units of volume

The two units of volume for solids in the metric system are the cubic centimetre and the cubic metre. For many purposes the cm^3 is very small, and the m^3 is very large, and intermediate volumes are needed.

One is frequently used in building, for example, the so-called 'cubic litre'. A small DIY concrete mixer is said to mix 42 to 84 cubic litres of concrete, and the wheelbarrow to go with it can contain 84 to 112 cubic litres of sand or cement, etc.

One can see that for such purposes both recognized metric volume units are unsuitable, and one can also visualize the intermediate 'cubic litre' invented for the special purpose.

Cars and volumes

Three main volume measurements concern cars.

Cubic capacity

This is the space for carrying luggage—also called 'cargo' capacity or loadspace. This is estimated with the backs of the back seats down flat. Probably because of the large size of the cubic metre, and the small size of the cubic centimetre, the space in the back of the car is given in cubic feet; this does not mean that car manufacturers have not got around to metric measurements—they use them elsewhere on the same cars. Examples of loadspace are: Austin Morris Maestro 50 cu ft, Austin Rover Ambassador 54.7 cu ft; these are volumes of air, of course. A good way to model these volumes is to collect and build up blocks of firm cardboard boxes, measuring one and calculating the number needed to match the car carrying capacity.

Fuel tank capacity

This refers to the maximum volume of petrol the car can take. 'Fill her up' means for the Maestro 11¾ gallons or 53 litres; the makers give both Imperial and metric volumes, as do the filling stations. For the Ambassador the volume of petrol is 16 gallons/72.7 litres. Some drivers will ask for one, and some for the other.

For most people, a display of what these volumes look like is quite striking; with some persistence and a little help from one's friends one can collect and line up 53 (or 73) 1 litre clear plastic drinks bottles, fill them all (or all but the last of the 73) with coloured water (to show that they are full), and label the collection—'this is the volume which goes into the . . . tank' (and will take you how many miles?). For comparison, the volume for the little Mini is 7.5 gallons/34 litres.

Engine size

The third volume connected with cars is the 'prestige' measurement, and probably the least understood: for the Mini it is 998 cubic centimetres, for the Maestro 1.3 or 1.6 litres, and for the Rover Vitesse 3528 cubic centimetres (called 3.5 litres).

Motorbike enthusiasts will recognize the difference between a 350 cc engine and a 750 cc. What is the top limit for an L rider?

This volume in each case is the total volume of petrol vapour and air mixture inside the cylinders—4 cylinders in the Mini up to 8 in the Rover Vitesse. This is, of course, where the power comes from, and where the petrol goes. It is often just called the 'engine size'. The range of volumes again makes an interesting and enlightening display. It could be extended by using data for other makes.

Length

Associated vocabulary

The concept of length brings with it a tremendous vocabulary. It is useful to collect some of the words associated with measurements of length:

how long? (excluding the time meaning); what length? how far along?

how far distant? what distance? how much farther/ further? a long way; very near, nearly there; halfway there

how short? a short length (!); a short distance

how wide? what width? how broad? what breadth? how far across?

how thick? what thickness? how far through? how thin? how narrow?

how high? what height? how tall? how far up? how far to the top?

how deep? what depth? how far down? halfway down/halfway up (!); how far across (for more or less circular or cylindrical objects); what diameter? (halfway across—radius?)

how far round? (perimeter *or* circumference, depending on shape. . .)
how far round the edge? (perimeter)

Adult native English speakers easily recognize and distinguish these quantities, but one must realize how large a number of words is involved before actual questions of measurement arise.

For more advanced students there is a further point to be noted. A distance in any undefined direction is a scalar, e.g. 1 cm, 1 km, while a distance in a specified direction is a vector, e.g. 2 m vertically upwards, 5 km due North. Such a distance in a straight line from A to B is called the displacement. Five times round the running track may be a distance of 1 000 m, or 1 km, but if the finishing line is at the same place as the start the total displacement is 0 metres.

Measuring instruments and units

Methods of measuring length, and the apparatus used, vary with the situation and the general size (or magnitude) of the object to be measured. Measurements will be considered here in three groups, those which can be called classroom and playground lengths, the very small lengths— roughly 1cm and under, and the longer distances, some of which cannot actually be measured, but may be estimated, or may simply be 'looked up' for interest.

Units for classroom and playground lengths

The correct and useful units here are the metre (m), centimetre (cm), and perhaps (although this is not recommended) the decimetre (dm) 10 cm long.

Instruments and tools

Of the measuring instruments or tools needed, rulers, metre and half-metre sticks, tape-measures, surveyors' measuring tapes, centimetre squared paper and transparent plastic grid should all be familiar. Others will be introduced for special purposes.

When measuring lengths the size of a playground, paces can give a very rough idea, but not only are people's paces different, but one person can make a considerable difference intentionally. These two sources of inaccuracy are interesting to test.

Metre-sticks, ropes marked in metres, the 'click wheel' with a circumference of 1 metre, and surveyors' tapes can all be used for measurements accurate to a metre. The mathematics of the 'click wheel' is not always understood by the users, who may just be counting the clicks.

Rulers are the most common measuring tools, but most of them have a serious disadvantage—the unmarked 'extra bit at the end'. Of course it is supposed to protect the scale, but how can one measure the tread of a step from front to back, or the depth of a shelf in a bookcase from back to front with one of these rulers? Some dead-length rulers, genuinely starting at zero like a number line, are essential.

Accuracy with a ruler depends on several factors—the markings may be rather wide, the ruler may be thick (so that one has to tilt the ruler to get the markings to meet the object being measured), the very small (millimetre) divisions may be confusing. On the other hand, if the ruler is only marked in centimetres, many measurements will fall between marks. Human judgement is far more accurate at a marking than between markings; we can take 'the nearest', and say so, or we can estimate. We are then better at 'guessing' halves and quarters than tenths—so in a decimal system the answer is probably to find a more finely divided ruler.

Practice helps, however, and can be consciously

carried out, e.g. when drawing columns, grids, and when folding paper to make strips. . . All kinds of diagrams, and even rubbings can be done on centimetre squared paper, giving familiarity with the unit most commonly needed.

The engineers' steel rule is excellent for accurate work since it is dead ended and perfectly straight, whereas wooden rulers may warp, and it is also finely marked in millimetres. The pocket 'pull-push' steel tape has special advantages because of the slight 'hook' on the zero end, which makes it possible for one person to measure longer lengths than he or she could reach. The linen or plastic tape with its metal end, and the 'wind-up' tape-measure are both very flexible. In addition, the 10 to 30-metre field tape has a ring, included in the measuring length, by which it can be hooked over a nail or peg while it is unrolled.

Measurement practice

Objects for measurement practice and investigation depend on the work in hand, but the more advance preparation and therefore familiarity with the methods, tools and units the quicker and more reliable the results.

Paper and pencils

How big is A4 paper?—a measurement that is often useful. What do you get when you fold it in halves? (A5, of course). And in quarters? What does A7 sized paper look like? Try it, and A8 . . . A10 is a struggle!

Which is bigger, A10 or a postage stamp? And how big is a standard stamp? This matters to collectors with albums (or at least to the makers of the albums).

How long is an ordinary new pencil? And how much goes each time it is sharpened? How short is the shortest usable stub of pencil?

Staples

How long are the two common sizes of staple—the little 'Bambi' and the standard ones? There are very large staples used on cardboard boxes—how long are the largest available?

Envelopes

What are the common envelope sizes? How *big* can a posted parcel be? This is a peculiar measure as it combines the length and girth!

Anyone with a handy horse might like to measure the horse's girth too. The standard measurement for horses is still height in 'hands' from the ground at the shoulder; a man's hand is assumed to have been 4 inches (i.e. 10 cm) wide (in fist form?) when this was established.

Film and slides

Photographic film, cine film and projection slides come in international standard sizes, stated in millimetres; 35 mil is a well-known phrase, the width of cinema film as well as camera film.

Kodak say that the camera exposure area measures 24 × 36 mm, the photograph, of course, being lengthwise on the roll. Projection slides need exact measurements for the card or plastic mounts, i.e. '2 by 2' (inches) or 5 cm × 5 cm, especially with an automatic slide changer.

The dimensions of the picture (image) which these slides will produce on the screen (or wall) are enormous by comparison (see page 31).

It is worth testing the accuracy of measurements made with a ruler marked only in centimetres, when the millimetres in between have to be guessed, compared with the same measurements

film and slide

made with a ruler marked in millimetres. Are measurements of things such as envelopes, stamps or slides made by putting them on squared paper or a plastic grid more or less accurate?

Lengths around the home

Useful measurements, as well as useful practice, can be obtained by organizing a round-the-home safari with a tape-measure or ruler, marked in centimetres, of course.

How wide is the front door? Are all the doors the same width? How wide is the narrowest one? Would anyone find it difficult to get through? Would all the tables in the place go easily through the front door? How could they be got in if they are too wide? Measure to make sure. Armchairs are sometimes a problem too.

How long is a standard bed? And how long is a standard sheet for it? And a blanket? What about a duvet? Think where the extra sheet length goes. Some people find ordinary beds, sheets and blankets all too short. If such a person can be found, how tall is she (or probably he)? What about the width of a camp-bed, a folding 'Z-bed' or a studio couch? How does this compare with the width of an ordinary single bed (of which there are two sizes)? How wide is a double bed? Twice the width of a single one? And a 'king-size'? (Not all kings were fat. . .)

'How long is a piece of string?' is a very old trick question—but the questions 'How long is a loaf of bread?' and 'Will it go into the bread bin?' have sensible measurable answers, even if all different. A genuine problem arises here in deciding how to measure things with rounded ends or edges, like kitchen work-tops or chair seats. Does one measure the greatest dimensions, or the top?

These may be familiar, in theory at least, from tessellation exercises; 'How big are the cork wall tiles?' ('305 × 305 × 3 mm' says one maker), and the ceramic tiles for the bathroom? (15 × 15 cm on the pack).

Knitting patterns tend to give measurements in centimetres, with inch equivalents in brackets. Knitting is far more difficult to measure accurately than tiles; the pattern sometimes says the knitting should be measured 'unstretched', but often also

insists that the knitter makes a sample square first. Get a few volunteers to try this, with the same size needles (see page 32), the same yarn or wool, the same numbers of stitches and rows. Then measure—how widely do the samples differ?

Lengths out of doors

Straight line lengths

Out of doors there are many straight line lengths, some of which are interesting: e.g. across the pond. Where the surface is wet, probably one of the best methods is to use a plastic-covered nylon clothes line, if possible in a fluorescent colour. Rope takes too long to dry afterwards, and is difficult to see in grass. To make sure that the line is across the surface of the pond, floats made of bits of expanded polystyrene can be threaded on. One end of the clothes line usually has a fixed ring, and a mark can be made at the opposite edge; the distance can be measured later. How? Well, there is a true story of a class who put all their own rulers end to end, borrowed more from another class, but finally said they still couldn't measure the corridor because they hadn't any more rulers.

floats on the clothes line

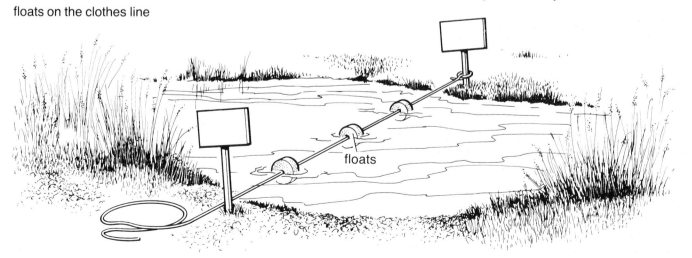

floats

Sports Day

Getting ready for Sports Day provides measuring activities, with rules (though rulers can be used here and there). Starting and finishing lines, for example, should strictly be 5 cm (2 inches) wide, and in chalk. The relay batons—smooth hollow tubes—should be 'not more than 30 cm and not less than 28 cm long' (for adults, of course). The long jump is 'measured to the nearest 1 cm' less than the distance from the front edge of the take-off board to the nearest touch. There is a fascination about the Olympic results and world records, but these can be daunting to read.

Do things (and people) look the same length standing up and lying down? This can be tested with metre measuring sticks fitted end to end—3 metres if possible. Look along the line on the ground, and then at the same length upright; many people find that they look different; is this the same problem as with tall jars and dumpy jars containing the same volume?

The high jump presents other small questions. The rule says that it is measured from the ground 'perpendicularly to the upper side of the cross-bar where it is lowest' (to the nearest 1 cm below the height measured).

Heights around the home

Start with the heights of the inhabitants, and the dog. . . How does the dog make itself taller, e.g. to see what's on the table? How much taller? Little dogs try so hard.

How high is the front-door? Is anyone around taller than that? And is there an old house nearby, where almost everybody has to mind their heads? How high is the ceiling? Do a guesstimate with a broom-handle, and then find a way to measure it (safely!). Is there a step-ladder? How high is it

from one step to the next? How many steps are there altogether? So how much taller does it make the user? How high are the staircase steps? This is where you must have a dead-length ruler or tape-measure. How many steps are there to go up one floor? How does this compare with the ceiling height? What height is the space inside a wardrobe? What would the tallest space in a wardrobe be needed for? How high is the space in a bookshelf for paperbacks? How high a space would be needed for encyclopaedias and atlases? What can you do if there isn't a shelf-space tall enough? How high is the table which is used for work, writing, breakfast. . ? How high are the seats of the chairs used at it? So how much room does that leave for knees underneath? All kinds of work are easier if the table and chairs are the right heights. But coffee tables are much lower—how much lower? And why does this seem right too? Some

houses and cottages in the country have marks on the side of the door-frame, with dates and initials against them. What would these mean? Do any of the initials fit the names of any of the children living there now?

Heights out of doors

Plants

Measuring the heights of such things as sunflowers and hollyhocks is relatively easy, using a metre-stick or tape-measure, even if the plants are taller than one can reach. More difficult to measure are the stem lengths of climbers; is one going to take the height reached, or the total length of stem?

Somewhat the same question arises with the heights ... top of the ...ss-like plants ...ing stage.

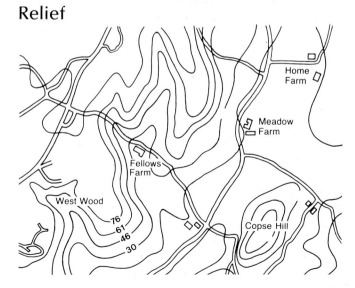

'Bounce'

Estimated heights are probably all one can hope for when testing the 'bounce' of different types of games ball, or the height of a trampoline jump. For both of these, a scale can be organized on the wall behind the event, with observers watching from as nearly the same height as possible (to avoid distortion due to looking from above or below). In some cases the courses of bricks in the wall can be used, the reachable courses being measured to establish a standard.

Tides

In the same way, the height of the tide at the shore can be found approximately if there are conspicuous and more or less vertical rocks, or perhaps the 'legs' of the pier. The difference between the sea-level at low and high tide is usually too great for any kind of measuring pole, e.g. spring tides: at Aberdeen 3.5 metres, Dover 5.7 metres and at Swansea 8.6 metres. On a sloping beach the rise can be observed, but only measured with difficulty.

Relief

Maps are available which show the measured heights of land, as in Ordnance Survey relief maps. Here the vertical interval (on the 1:50 000 maps) between contour lines is 50 feet (15.24 m) – and the heights of 'peaks' are given to the nearest metre above sea-level.

Bridges

Bridges over the road often have a notice on them saying how high the space under the bridge is. Look for some, and make a note of the heights. How high is the passengers' space inside a bus? And is it the same on the top deck of a double-decker? Is this high enough for the tallest girls and men passengers? And the conductor?

A goods vehicle over 13 metres long must have a notice on the back saying LONG VEHICLE. How does this compare with a bus or coach?

Depths

Shallow water

For relatively shallow water, or other liquid, where it is not possible to see through the side of the container, the dipstick method is a good one. Many wooden rulers not only have that unmarked bit on the end, but may warp if they stay wet. A firm plastic dead-end ruler, e.g. the type made by Taskmaster, with centimetre markings vertically on one side, is excellent. Osmiroid make a depth gauge, in which the ruler slides through a crossbar. This rests on the top of the container whose depth is being measured; the instrument would not be very practical for measuring the depth of liquids.

To use the dipstick method efficiently one needs either to get one's eye-level at the same height as the top of the liquid, or to use the car drivers' method. Start with a dry ruler, dip it vertically to the bottom of the liquid, take it out carefully, and see how far up it is wet! This works very well for the oil in the sump of a car—not the petrol, of course.

Deep water

In deeper depths, such as the deep end of the swimming baths, or a deep pond, or the sea off the end of the pier, one can devise a string version with a weight on the bottom end, and if necessary some kind of 'fishing rod' to get the top end far enough from the bank. If the string/rope is marked in advance with waterproof ink, the depth can be read off; if not, the wet length can be (roughly) measured as it is hauled up.

How far across?

There are two possibilities for measurement here—external and internal diameters, e.g. the stopper and the bottle-neck. External diameters of circular or more-or-less circular objects are very frequently needed in everyday life.

Rulers and calipers

Diameters that can be measured with a ruler include distances across drain-pipes, cut ends of hosepipes, cups, saucepans, buckets, etc. However, some of these will have handles, which may have to be considered, for example in planning shelf space. This applies also to scientific equipment such as beakers, test-tubes with rims, and funnels with handles. The maximum diameter is often the measurement which matters.

Bow calipers can be large plastic models intended for teaching, or steel ones for engineering use.

How wide are the cat's whiskers, actually? A ruler will be very hard to use here, but bow calipers give a good result when the distance between the

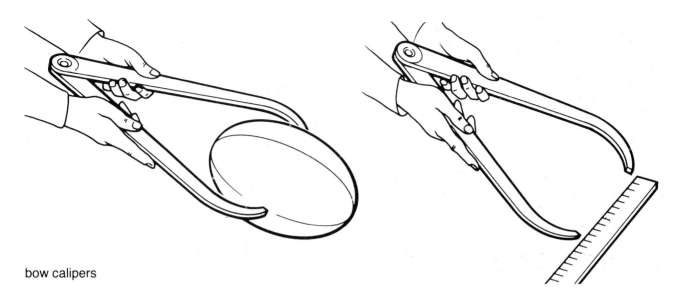

bow calipers

sliding or graduated calipers

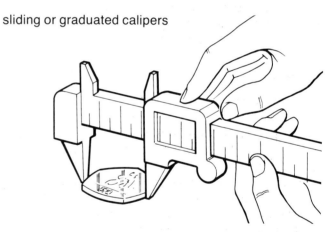

tips is transferred to a ruler. One tip of the calipers can be placed so that it touches the zero end of the dead-length ruler, and the position of the other tip read on the scale.

How wide is a racing bicycle tyre? How wide, for comparison, is a car tyre? Make sure first that there is nobody in the driver's seat, and that the car is parked.

The width of a crab shell, or the dimensions of a rugby football, could be much more difficult to measure without some form of calipers.

Sliding calipers (graduated calipers) are extremely valuable for accurate internal and external diameter measurement. The 'jaws' can be adjusted so that the sliding section and the fixed end projection just touch the two sides of the object, so that the dimension can be read off the scale.

For example, carry out a test with a 50 pence piece; although this has 7 angles round the edge, the slight curve opposite each was very exactly calculated so that the coin could not lodge in a

slot-meter. . . Test the newer 20p pieces as well. One coin, the 2p piece, was planned as a possible unit for measurement of length—what is its diameter? Charities at one time collected money by organizing a 'mile of pennies' (the old large ones). This might perhaps be revived—in a safe environment. But perhaps one would need a shorter length. How do the diameters of the current coins compare with their values? Do they vary in the same order? Their thicknesses can also be measured with good sliding calipers.

External and internal diameters

There are many examples in everyday life, in science and in engineering where an external diameter has to be correlated with an internal diameter. This is not always a matter of measurement—it may be simply a 'go/not go' situation, as with buttons and button-holes, or corks and bottle-necks. However, somebody somewhere measured them both. In engineering, for example in the bore of a car engine cylinder, an accurate fit is vital. In other cases testing for 'fit' shows variations from 'stick' through 'good fit' to 'wobble'. What the engineer calls 'tolerance' is the maximum difference between acceptable tightness to acceptable looseness of fit—remembering of course to leave a little space for oil! Consider some simple examples:

Rings

Very few people will measure the important finger with calipers, but the jewellers' shop has a 'ring stick' on which one slides a ring known to fit the finger, and looks at the marking where the ring stops. When ordering rings by post, one is sometimes given a line of circles, with numbers to match a known ring on (but these are not standardized). Between 'go' and 'not go' there are examples which could be called 'half-go', such as the ring on the ring stick.

ring sizes

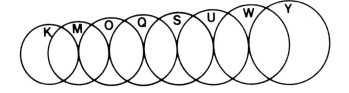

Corks and stoppers

What about the cork in a bottle? Will it go in at all once it has been drawn? Will a new cork from the shop fall right in? One needs three diameters altogether to be sure it's the right cork.

Ground glass stoppers in glass bottles are different again, and much more exacting; with small-scale internal calipers the diameter halfway down inside the neck can be measured, to be matched by the external diameter of the stopper.

Candles

Candles are more easily adjusted to their holders, providing they are wide enough to start with. Internal diameter may be measured directly with sliding calipers or the distance between the tips of bow calipers can be transferred to a ruler.

Funnels

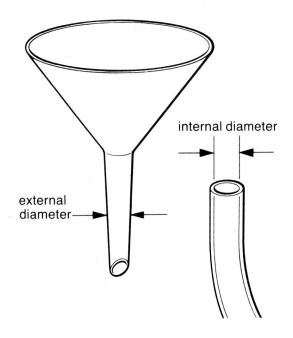

A common problem in fitting up apparatus is posed by the question 'Will this funnel fit into that plastic tubing?' Or in dry weather 'Will this sprinkler fit my hosepipe?' Here we need one external diameter and one internal diameter, for which a good ruler or steel rule laid across the ends of the relevant tubes may be adequate, but if the ends are cut off diagonally, then a sliding caliper is needed.

'Piggy-banks' and letter boxes

Most money-boxes are supposed to take any coins you like to put in, so how long and how wide must the slit be? Most money boxes were made before the arrival of the 'round pound' coins. Will the new £1 coin go in? It is not very safe to trust to the 'go/no-go' method here. Careful measurement would be better than getting it stuck.

Simple testing on the spot is the normal method for another set of common apertures—letter boxes.

Will the letter/packet go in, or not? However, a different dimension is available here, the diagonal internal dimension. The Open University requests its students to fit a letter box which will take A4 packets without folding (though these are surprisingly difficult to buy). Alternatively there may be enough space under the door. . .

Some road-side pillar-boxes have remarkably small openings too. How would one set about a survey of suitable letter box slit sizes? Perhaps with bow calipers? Or simply with stiff card 'gauges' made to the size one considers the slit should be able to take, e.g. A4 with envelope?

Marbles and drains

Every now and again games of marbles become popular in playgrounds and on quiet road pavements—but marbles roll down predictable slopes leading to drains. This is another 'go/no go'

question, best not solved by direct testing. Sliding calipers give the diameters of the marbles in play—but the width of the gaps between drain bars is variable, and probably best measured with a V-shaped card 'width gauge', or the Osmiroid type plastic version. In any case, to be effective the gauge needs to be slid along each aperture from end to end, to detect the width of the widest part of the openings. Comparison with the marble diameters prevents disappointment.

Making card sliding calipers

This is interesting and economical to do. One type is made on a ruler, which holds it stiff, but this type has a built-in disadvantage—it can't start from

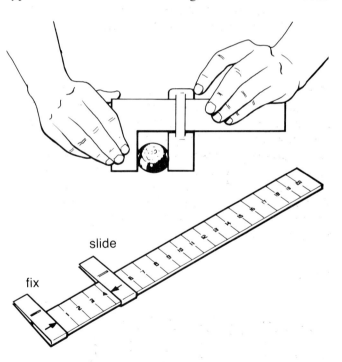

slide

fix

0 (zero) because you need the end of the ruler to fix the 'jaw' which does not move. However, for some measurements, e.g. the heights of seedlings from ground level, one can simply make one movable 'jaw' to slide along the ruler, and 'zero' can be the beginning of the dead-length ruler. The other all-card type of DIY calipers has a cut-out fixed jaw at the left-hand end, a stick-on or hand-drawn scale beginning from the inside edge of this jaw, and a sliding jaw made from card with two slits in it, to push along the scale (ruler). This type is useful for measuring diameters of seeds from coconuts to conkers or avocado 'stones', but is not reliable enough for very small objects, since there is too much play between scale and slider. For tree-trunk diameters one can use a metre-rule and two large sliders.

Diameter grading

Grading by a kind of diameter gauge takes place very frequently in industry, as well as in some areas of science. A grid, a sieve or a riddle has definite sized openings, allowing smaller particles or objects through and retaining the larger ones. The dimensions of the two classes are often quite specific, though if the material contains some round 'fat' items and some long slim ones, the criteria may be very difficult to define. This may happen with materials as diverse as lumps of coal, potatoes, or the millions of under-2-millimetres creatures in the sand at the sea's edge (the meiofauna). Of these last, the shrimp-shaped animals will stick where much longer worms slide through the 2 mm mesh of the nets used in research. Pouring solid particles through an opening raises a different question. It is said (how reliably?) that with common materials such as breakfast flakes the opening in the packet needs to be more than six times the maximum diameter of the particles; if not, they tend to 'bridge across' and block the flow. The blocking phenomenon is fairly well known—and might be worth testing.

Curved and zigzag distances

These are much more difficult to estimate by eye than straight lines. Try this kind of test: a centimetre is this —— long; 100 of these, end to end, measure 1 metre. But how long is this line from 'Here' to 'There'? And how does one find out? Perhaps by 'stepping out' centimetres with a

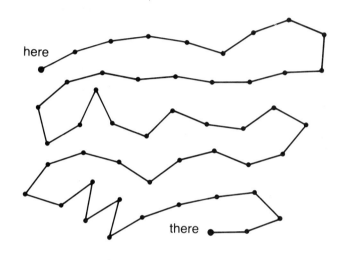

pair of compasses? In the same sort of way one can divide a circle into 6 by using the radius—though this is not useful for measuring the circumference, of course.

A good exercise can be built on the straight/wavy length problem by taking a number of equal long strips of thin card or stiff paper; **1**. is kept straight, **2**. is coiled, **3**. is curved into a wavy form, **4**. is folded into a coarse zigzag, **5**. into a fine zigzag, and **6**. into halves, quarters, eighths. As many 'outsiders' as possible are then asked to look at the *top edges* and to place the strips in order of length. The same type of problem is provided by the old catch question 'How long is a piece of string?' A collage of string, e.g. tight coil, loose coil, wavy line, maze-type patterns, a square and a circle—each made from the same length of identical string—is convincing evidence that guessing is not good enough. It is sometimes suggested that one should match a piece of cotton to a twisting line, e.g. a leaf-mining insect's tunnel in a leaf, then measure the cotton. This is easier to say than to do.

Measuring all the way round

Here the measurements are of circumferences and perimeters. Distance round is perceived as different from the straight-line measurements which one can make with a ruler. For many situations the tape-measure is most satisfactory—for tree-trunks, marrows or human waists. How far is the old country saying 'Twice round your wrist, once round your neck; twice round your neck, once round your waist; twice round your waist, once round . . . a haystack' justified?

Tree-girths

The tape-measure gives a direct measurement, but of course string can be used and measured afterwards, especially if the object is very large. Tree-girths are frequently measured, for various reasons, e.g. to compare with other trees, to find out how much one particular tree grows from year to year, to estimate the volume of wood in the trunk, or perhaps for the *Guinness Book of Records*—the largest tree in Great Britain is an oak in North Wales—more than 43 feet round. . . But how high up does one measure the girth of a tree? Is it going to make a lot of difference? The *Guinness Book* says at 5 feet up, other books say at 1.3 metres (above ground level, on the uphill side of the tree!) 'As far up as you can reach' is a non-starter, since the results cannot be compared with those of other people, nor perhaps with results by the same experimenter next year.

Belts

Belts of all kinds are good material for circumference measurements, making an allowance for overlap or joins; for example, belts round waists, the 'open' belt round two wheels such as the fan belt in a vacuum cleaner or car, and the bicycle chain. There are also belts which cross over, making the driven wheel turn the opposite way to the driving wheel—these crossed belts are common in engineering, and models are easily made with cotton-reels and rubber bands; the belt has to be long enough to allow for the cross-over.

Trundle-wheels and bicycle wheels

Trundle-wheels use one circumference (that of the wheel) to measure another circumference or perimeter when the playground or running-track is the object of the exercise. The 'click' indicates '1 metre' distance, but the exact connection is often not clear to the operator—since not everybody actually measures the click-wheel before going into action with it. A tape-measure round the rim is a practical way to start.
A bicycle wheel makes an excellent alternative. With a very chalky chalk mark on the tyre it is possible to mark out a distance simply by riding it—and the chalk can be renewed when it gets faint. The circumference of the wheel, including the tyre, may already be known, or can be measured with the usual tape-measure. This kind of field activity is best repeated for a check—or two cyclists can do the circuit at the same time, counting as they go. When the idea is clear, they may not need to use chalk, but can count the number of times a coloured marker on a front-wheel spoke comes to the top.

Lengths and light

These are measurements specially connected with mirrors and lenses.

Flat, or plane, mirrors

When you look at yourself in the mirror, where is the 'image' of yourself that you see? The first point is that it is not real—it doesn't exist. The rays of light going from you to the mirror have simply bounced—have been reflected—back again. *But* when this is tested by experiment, they *seem* to have come from behind the mirror, which is obviously impossible, as no light can come through a mirror. However, it is perfectly possible to measure where the so-called 'virtual' (not-real) image seems to be.

(a) Walking towards a long mirror or glass door with a dark background, one can see one's image becoming taller and taller, just as if one were moving towards a real object (oneself).

(b) If a small mirror is stood up vertically on lined paper, any mark made on a line in front of the mirror can be seen to have a (virtual) image in the mirror, and the number of writing lines from mark to mirror can be counted; this can be seen to have an equal counterpart from the mirror to the image, which appears to be an equal number of lines behind the mirror.

'image' in the mirror

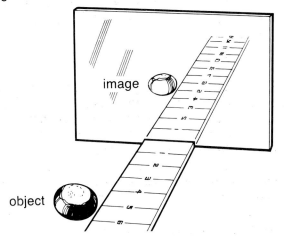

image

object

(c) With a plastic mirror, which has the reflecting surface on the front, and a dead-length ruler, exact distances can be measured. The ruler is placed at right angles to the mirror surface, and an object is stood either on it or beside it. The position along the ruler is front of the mirror can be checked with the distance along the image of the ruler 'behind' the mirror. The two appear to be equal.

Convex (magnifier) lenses

A convex lens bends the rays of light which go through it so that they go to, and through, a single point in space—the focus. (This applies to heat rays as well as light rays, hence the name—Latin for 'hearth').

(a) The focal length, which depends on how curved the lens surfaces are, can be found by making the brightest, hottest, spot on a sheet of paper by holding the lens between it and the sun. The distance from the centre of the lens to the hot spot is the focal length.

(b) Holding a convex lens up between a window and a vertical sheet of paper gives an image of the window on the paper. This image is upside down; the distance between it, when very sharp, and the centre of the lens is approximately the focal length. In both of these examples the image is actually on the paper, and is therefore a real image; the rays of light, and of heat, really go there. (And, of course, the image of the sun in the first example is upside down too—though how could one know!)

(c) A slide projector has more than one convex lens inside it; however, the resultant image—which can be caught on the screen, and is therefore a real image—is the opposite way up to the slide (as many amateur projectionists have discovered by trial and error). The interesting measurement in this case is the magnification which can be obtained on the screen. This begins with accurate measurement of the film as it appears in the slide—smaller than the nominal 35 mm of the film itself in width, and about 36 mm in length. Then with a good sharp image on the screen, the 'picture' can be measured in both directions. The enlargement (magnification) can be very great indeed. It is worth finding the distance from the projector to the screen, and altering this to find the effect on the image size. For some study purposes an image of about A4 size projected on a sheet of paper held vertically can be of great help, and this distance is worth recording for future use.

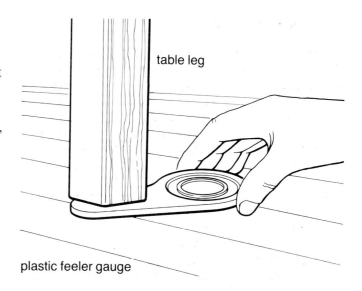

table leg

plastic feeler gauge

Very short lengths, widths, diameters, etc.

Small dimensions often need specialized measuring tools; here are some examples:

Narrow spaces

A table wobbles, or a chair rocks (though it is not a rocking chair). How big is the space under the short leg? And what can one do about it? One can use trial and error, by folding paper, folding it again, or even using a beer-mat? The narrow space can also be gauged with a feeler gauge. The plastic set from Osmiroid contains thin plates of definite thicknesses—1 mm to 5 mm; one, or two together, can be slipped into the space to find which, or what combination, just fills the gap. This gives the thickness of the material (in engineering, a shim) needed. Other situations for this kind of gauge would be the space which allows a draught under the door, or a gap between window and frame.

The much thinner steel feeler gauge set is, of course, for testing the spark gap in spark plugs.

Widths between writing lines

One space between 'rulings' can be measured with an accurate ruler marked in mm, but it is much more satisfactory to measure a number—say 10—spaces, and then to divide the result by 10. In this way, any inaccuracy is also divided by 10; the method is a general one for small measurements.

Heights of letters

Letters or characters in typing and printing can be measured as well as the spaces between the lines of type. Printers have special measures, but it is important for the ordinary typewriter user to know such things as the number of characters in a full line, the difference between single and double spacing, and the differences between the sizes of the actual letters on different typewriters; some are much easier to read than others. All of these are worth measuring and checking.

Thickness of very thin things

One can measure, for example, sheets of paper: two common thicknesses of typing paper are sold by weight and in 500 sheet (ream) boxes. The weights are 70 g/m² and 85 g/m², but we do not very often buy square metres of paper for typing.

The thicknesses of the two grades (called 'substance' on the box) can easily be compared by measuring the whole 500 sheets in a new packet of A4 size paper.

Thicknesses of needles

The thickness of knitting needles can be very confusing unless one has some kind of gauge or measuring tool!
The English sizes are the old engineers' SWG (standard wire gauge) sizes, and the thinner the needle the larger the gauge number, e.g. 14 SWG, size 14 knitting needle, is 2 mm in diameter; 6 SWG, size 6 needle is 5 mm in diameter. The American standards are different: English 6 is American 7 and measures 5 mm; English 14 is American 00 and measures 2 mm. Knitters need some way to measure. A rough test is to push one needle through stiff paper, and then try another through the hole; a smooth fit shows the same size. . . This is an engineering kind of 'go/not go' test. The accurate measurement is made with a sliding caliper or with a screw gauge.
The sizes for sewing needles increase from 1 to 12 or 15 as the needles get thinner; very thin, very long, needles (10 to 15s) are made specially for threading beads. A micrometer screw gauge would be the only tool for measuring such diameters, though they could well be compared, as can single hairs, using a low power of the microscope. Special microscope slides are made with a graticule or scale marked in tenths of a millimetre for this kind of work, and for measuring small organisms such as amoeba, water fleas, and real fleas too.

micrometer screw gauge

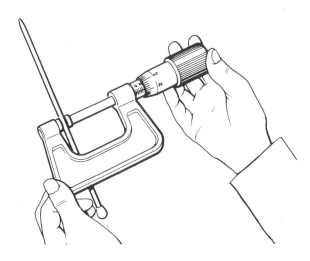

Very shallow depths

The rainfall in Britain is too small to measure accurately by direct use of a dipstick. Some rain gauges have a wide funnel-like top, leading into a much narrower calibrated container at the bottom. This makes a little rainwater go a long way up the scale, which gives the correct depth of rainfall in millimetres and decimal fractions of an inch.

Very long lengths, heights, distances, etc.

Playgrounds, playing fields, sports grounds and streets provide many occasions for useful and often interesting measurements of distances. Paces, trundle-wheels, metre-rods, surveyors' tapes can all be used for rough or more accurate length measurements, and to measure out required distances (a different project). Apart from necessary measurements, for marking out pitches and courts or for judging throws, etc., some known values can be much better comprehended if they are made concrete on the playground, e.g. the length of the Boeing 747–200 (70 metres), and the wing-span (60 metres). The height at the tail (19 metres) may go up the walls of a multi-storey school, but with a bungalow-type building one can only say how many times higher the plane would be. Would a Concorde go into the school playground? How would its length look along the road outside?

Distances along roads

Some bicycles have a distance meter on the wheel; this is comparable with the trip meter in a car speedometer, and gives the distance travelled. The mileometer gives (unless it has been turned back) the total distance covered—the 'miles on the clock' of the car salesman. There are still a lot of miles around in Great Britain, though road signs tend to say 'Max. 30 mph/Max. 50 km/h' giving both units. Most speedometers give both units, though some now only have the metric scale. This has to be got used to. It is worth while finding out on a good map how far a kilometre radius stretches from, say, the school gates—more than ½ mile.

Other good work can be done with Ordnance Survey maps, scale 1:50 000. Two types of measurement need to be considered: the 'crow flies' distances, and the distances by road or rail, which may be very much longer. The little map-measurer, in principle like a tiny trundle wheel, is valuable once the scales have been mastered, though this

can be tricky. The *AA Members' Handbook* has a Mileage Chart which can be used to check road distances as found on the map—and also for occasional interest figures, e.g. London to John O'Groats (705 miles), London to Land's End (301 miles), while the notice actually at John O'Groats says 'Land's End 874 miles'. Using an atlas and air-lines' maps one gets some other surprising results, partly because of the impossibility of flattening the surface of the world on to paper. However, try comparing the distance from London to the north of Scotland with that from London to Berlin. And the non-stop flight London to Vancouver (4731 miles, 7614 km) with the range of the Boeing 747–200 (6 000 miles/9 650 km); no problem!

High heights

These range from those we can actually measure, e.g. the tallest giraffe (George, in Chester Zoo was 6.09 metres, in 1980), a male ostrich (nearly 2.5 metres), to heights we can only estimate, such as the heights of tall trees (the tallest is *Sequoia sempervirens*, 112.10 metres—how far would that stretch if it fell in the road outside the school?), the London 'Monument' to the Great Fire (only 61.57 metres), and Cleopatra's Needle (21.2 metres), which incidentally had no connection with Cleopatra. The height of the Monument can be partially measured, in terms of step-height and number . . . but there is quite a lot more on top.

Area

Area is not very easy to describe: the area of a tabletop is how much space there is on it, the area of the surface of an orange is how much skin there is on the outside.

Comparison of areas

Look into a very small room, say a store-room, and then into a large room or the school hall. The area of floor (often called floor-space) will make an impressive comparison.

Units of measurement

What units are available? At early stages, arbitrary square units such as the actual carpet tiles on the floor, the smaller tiles on a wall can be used. Since these are of standard sizes, the areas can be converted into measurements; carpet tiles are often 40 × 40 cm, wall cork tiles 30 × 30 cm.

For measurements, the units will normally be square centimetres, or for larger areas square metres. Millimetre squares are so small that counting is very tiresome, and would only be useful in any case with minute objects.

Any amount of work using centimetre-squared paper is valuable, e.g. designing mosaics, copying calculator keyboards (7 8 9 along the top) and modern telephone keypads (1 2 3 sensibly at the top), and making all kinds of graphics. The 'feel', the 'look', of a square centimetre becomes a matter of course.

Measurement of areas

This can often be done best simply by counting the centimetre squares covered, and even if the area is rectangular it is no bad idea to check the calculation this way (see APU reports).

The usual rule for irregular shapes is:

(a) count all whole squares in the shape as whole squares (of course);
(b) count all more or less exactly half-squares as halves;
(c) count parts larger than half as wholes;
(d) don't count parts smaller than half at all.

The result is a good approximation, since the last two groups cancel one another out.

3 cm square

dry peas   soaked peas

A simple, flat leaf

To find the area of a leaf hold it down firmly on a sheet of paper ruled in square centimetres, draw carefully round it, and count the squares and parts. There are many extensions to this, e.g. finding the largest available (fallen) leaf in the autumn; finding the actual area of surface of a leaf from which transpiration could take place (mainly the under surface for most leaves, of course); measuring the areas of green and non-green parts of variegated leaves.

Two points need care: the fallen leaf is specifically mentioned so that a class does not enthusiastically strip the lower branches of the perhaps only tree in the playground; a second point is that in the making of a block-graph of the areas of leaves on a branch, there may be far more small leaves than would give a balanced graph of normal distribution—simply because young leaves may be included before they have had time to grow to their full size.

The growth of young leaves

The leaves of a silver birch or a sunflower, for example, can be very satisfactorily measured by first marking the leaf with a plastic-bag closure twisted loosely round its stalk, and then drawing its outline on squared paper held behind it against a sheet of card—repeating after say a week. By using centimetre-squared plastic (OHP 'acetate') called 'metric area measuring grid' against the leaf, the area can be counted on the spot or drawn with water-soluble OHP ink.

A hand

A more difficult exercise of the same kind is to find the area of one's hand—the one not used for drawing. The first point to decide is where the hand ends—a limit can be established by discussion. Next—whether the fingers are to be together or spread; because of the difficulty of drawing round fingers, the results will be different. However, this can be made into a good demonstration of the

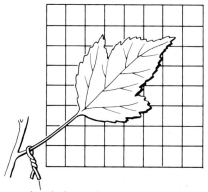

plastic bag closure

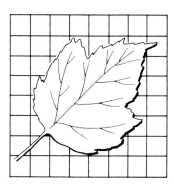

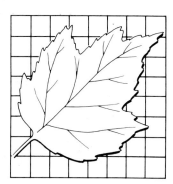

need for a standard method; two drawings are made with the hand in the same place and position—but for one, the pencil will be kept vertical all the way round, and for the other the pencil will be sloped so that the point goes under the curves at the edges. The resulting areas for the same hand will show what a difference the method can make.

Importance of areas

Areas are important in a wide range of circumstances: for other examples consider the following:

Evaporation

The effect of area of surface of a liquid on the rate at which it will dry up can be demonstrated by pouring the same measured volume of water into two containers, one wide and flat, the other tall and narrow. Both are left in the same conditions and either the volume decrease in a certain time can be measured, or the volume left in one when the other has just dried up. This kind of result has many consequences in science as in everyday life.

The area of surface is often circular, and counting the centimetre squares which cover it will for many be much easier and more convincing than calculating it from the formula.

Pressure of feet on floors

The area of surface under shoe-heels can be measured and hence the pressure when the wearer puts his or her weight on one heel on say a soft lino floor can be calculated. The entire weight of a hefty young woman on one (or even two) heels not much larger than a square centimetre can make quite an impression. Compare this with the spread load on the area of an average male shoe-heel, the wearers being of equal weight.

Light and heat gain and loss

Windows of rooms, sheds and greenhouses facing in the same direction and with the same type of glass, will show the importance of area in the amount of light, and in different seasons, heat gain or heat loss. Most schools can provide the experimental conditions for such tests.

Tests of absorbency

The weight of water which different materials can soak up—an important factor in everyday life as well in science—can be investigated using equal areas of fabric, paper, towel, etc. The best way to record the results will be with reference to a standard area, e.g. 'per 100 cm^2' (a square 10 cm × 10 cm is convenient for experiments too).

Area of a circle

The area of pastry for circular jam-tarts may begin as a problem in home economics—but the area of a circle is important, and difficult. The jam-tart

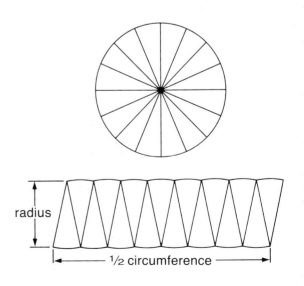

question brings in tessellation again—how does one cut out the maximum number of circles with the least area 'left over'.

The actual estimation of the area of a circle can be approached by cutting up a circle of paper (or of tart pastry) into narrow sectors, and fitting these wedge-shaped pieces alternately point upwards and point downwards, close together, making an 'oblong' shape. The slightly wavy top and bottom makes very little difference to the result—which can be placed on squared paper, or simply measured. The height of the shape is almost the radius of the circle, while the circumference is shared equally by the top and the bottom (i.e., width is half circumference). This can be checked by putting the complete circle (or its twin) on squared paper and again counting squares—though with some problems over the part-squares all the way round.

Area of surface of solids
Rectangular boxes

The area of surface of rectangular boxes is, of course, easy to measure—either by opening the box out flat, or by taking each side separately (the lid is surprisingly often forgotten).

Spheres

The real problem of area of surface of a sphere has been faced by map-makers throughout the centuries, and the common solution in today's atlases makes countries like Greenland appear vastly larger than they really are on the curved surface of the earth. However, a model can be attempted by finding a spherical and thin-skinned orange, cutting through the skin longitudinally, say 8 times, peeling off the sections and flattening them on a centimetre squared grid (perhaps plastic/acetate rather than paper?) for

measurement. If a cut is made round the 'equator' of the orange before the skin is peeled off, the triangular sections can be arranged head to tail and as closely as possible, as in the circle exercise.

Irregular objects

The area of surface of extremely irregular objects such as ourselves is of great importance, e.g. in medical science. The casualty department of a hospital frequently has to estimate the area of burns, and the proportion of the total skin area. A small area can be treated in out-patients; a medium area demands in-patient care; a large area needs the attention of the intensive care unit. (The depth of the burn is also vital information, of course).

Exact measurement of skin surface area is impossible, since skin can be stretched, and since there are many small folds. A rough, but still useful, estimate can be made by considering the body and limbs as cylinders, wrapping each neatly in paper (newspaper?) so that the edges of the tubes just meet, and finding the areas of the sections. The head can be taken either as a sphere or as an extension of the neck.

The same method can be used to find the area of surface of an animal, so long as the animal is a co-operative one. The dimensions of a dog or cat are easily found—with a soft tape-measure rather than by wrapping it in paper. Then the cylinders can be constructed without the inhabitant.

A rough but reasonable estimate can be made for small compact animals such as a mouse and a guinea-pig as follows: first get the 'feel' of the volume of each animal as it would be contained in the hands. The weight would be useful as well, but is not essential. Then select, at a green-grocer's or supermarket, a potato as nearly the same size and shape as the curled-up animal would have in one's hands. Peel the potato just thickly enough to be able to spread out the peelings on squared paper, and count the squares in the total area.

Surface area and volume

The area of skin surface on the outside of a warm-blooded animal is important because this is where heat is lost. The elephant waving those enormous ears is neither listening nor threatening with them—just cooling itself. One could get a fair estimate of the size without borrowing an elephant, but using a good photograph and easily obtained elephant statistics.

However, heat loss can be a question of life or death for small creatures, whether mice or human babies, in cold conditions. The amount of heat lost by a large creature—horse or man—may make very little difference to the temperature of the creature; the loss of even a little heat may affect the survival of a small one. The factor which makes the difference is the ratio of surface area to total volume ('size'). This ratio is a difficult concept to grasp, and may need a number of different examples. The basic fact is that the larger the object—block, box, potato or person—the less skin surface it has in proportion to its size (volume). This is best tested with easily measured cubes.

Try it with cubes such as Centicubes, building up cubical blocks, or with cubical boxes made of squared paper (for ease of finding surface area.)

The surface area does not get bigger at the same rate as the cubical blocks of cubes. Start with 1 cubic centimetre, then build a 2 cm cube, then a 3 cm cube, then a 4 cm cube, and so on. For each size, count the number of square centimetres on the outside, and the number of cubes in the block (solid, of course). The 3 cm cube will have 27 small cubes in it, but does it have 27 times as much outside surface as the single Centicube? This would be 27 × 6, or 162 square centimetres.

1. volume 1 cc
   surface area 6 sq cm
   ratio 1:6

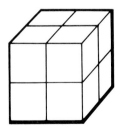

2. volume 8 cc
   surface area 24 sq cm
   ratio 1:3

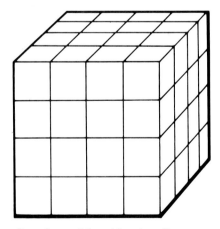

3. volume 64 cc (4 × 4 × 4)
   surface area 96 sq cm (6 × 16)
   ratio 1:5

Here is the main reason why babies need to be better wrapped up than their parents when they go out of doors in very cold weather. The smaller they are, the faster they lose heat.

Large areas

Areas of 1 square metre are often taken for the study of plants and small animals in field-work.

These usually need to be chosen at random, so that they are not picked out for any exceptional population. The boundaries are set either by dropping a frame made from 4 metre–sticks fixed into the form of a square, or by using strings or tapes. If 1 sq m is too large, square frames 50 cm along each edge can be used (with care that the results are multiplied by 4 to get sq m numbers). The squares are called quadrats.

Field dimensions are in decimal usage given in hectares (ha). One hectare measures 100 m each side, so this is quite easily marked out on an ordinary playing field; 100 m is only about 8 yards longer than the old 100 yards sprint track.

Hectares are normally used in agricultural science, e.g. for crop yields, since acres are useless for international comparisons.

Mass and weight

Basic points

1. Every object or material (peanut, potato, petrol or planet) has mass.

2. The mass of an object (or quantity of material) is found by balancing it against man-made standard masses.

3. The units of mass used in science are gram (g), kilogram (kg), and tonne. 1kg = 1 000g; 1 tonne = 1 000 kg.

4. The mass of an object is the amount of 'matter' (substance, stuff, material) in it. This stays the same whether the object is on the earth, on the moon, on its way from one to the other, or out in space.

5. All masses attract one another, but this can only be noticed when at least one of the masses is enormous.

6. The attraction between the sun and the planets (earth, Jupiter—mass 318 times that of the earth, etc.) holds them in orbit, so that they do not fly off into space.

7. Similarly the attraction between the earth and the moon keeps the moon circling as the earth's satellite.

8. This attraction is the force called gravity. The force of gravity pulls objects and materials towards the centre of the earth, and gives them their *weight*.

9. The force of gravity due to the earth decreases with increasing distance from the earth's centre, though it never becomes zero. What is often called 'weightlessness' is only apparent weightlessness, the effect of gravity being neutralised by other factors.

10. The less the mass of the large attracting object, the less its gravitational force; the force of gravity on the moon, due to the moon itself, is only one-sixth of that on earth due to the earth, hence weights of objects on the moon are correspondingly less. Their *masses* remain the same wherever they are.

11. The correct units for measuring a force, e.g. gravity, are *newtons*. Some spring 'balances' (scales) are marked in newtons; 9.8 newtons are equal to the downward force exerted on a mass of 1 kilogram (which engineers call 1 kgf). The newton is unfamiliar to most people, and is made more acceptable by describing it as 'about the downward force exerted on an average apple'! (at very roughly 5 to the lb.)

12. The vocabulary in use for finding mass and weight is so muddled and confusing that probably the best thing to do is to use the terms which other people understand, being clear in one's own mind what is correct but not worrying about it. In ordinary activities the distinction will make no practical difference to the results.

Measuring weight

(a) Establish the 'feel' of the downward force of gravity, e.g. with different objects on the hand held out horizontally.

(b) Hang a rubber band on a firm hook or drawer-knob, and imitate the force of gravity by pulling downwards with a finger in the band.

(c) Fix a hook, e.g. a slightly opened paperclip, on the lower end of the band, and hang on it a small plastic bag or a fruit or nut net. Add small equal loads, e.g. marbles, and watch the stretching of the rubber band.

(d) Start again without the load, fixing a vertical strip of card behind the band, and marking the level of the top of the paperclip after adding each marble.

This small activity is a model of the process of calibrating a spring 'balance'—which should be called a spring scale, since it has no connection with balancing.

There are several useful variations on this model, e.g. using several rubber bands looped into one another, making a longer extensible strip; using two or more similar bands side by side, strengthening the 'spring'; if the support is strong enough, using a car or motorbike luggage cord (which has the hooks just ready).

(e) Repeat any one of these trial activities, adding standard 'weights' (which are technically masses, but which will continue to be called weights) one at a time, marking the scale for each addition, and thus producing an instrument which can be used for actual 'weighing'.

(f) Weigh objects with it.

simple spring scale

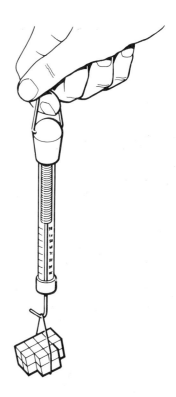

Notice the spacing of the markings on the scale—they will probably not be equal distances apart; rubber is not the best material for weighing 'machines', but it shows the method very well.

Scales and balances

Commercial spring scales

Common sources of inaccurate results include:

(a) the scale not hanging freely, but with either the case being held in a slanting position, or the load touching some object, e.g. the wall behind it;

41

(b) the spring having previously been stretched too far, so that the pointer does not return to zero. In this case, the unloaded position can be taken as the new zero, but one has to remember to allow for this at every weighing;

(c) the observer's eye-level not being directly opposite to the pointer (this is not particular to spring scales, of course).

Some spring scales have a dial instead of a linear scale; these can often be adjusted to get the pointer exactly on zero.

Kitchen scales

These usually have springs which are compressed rather than stretched, and a direct-reading scale marked in grams and kilograms. Some have two springs inside—a soft spring which gives readings for light objects, and a stiffer spring which takes over when the soft one has been fully compressed, and which gives readings for heavier objects.

Grams or newtons?

Because the force of gravity varies very little indeed on the earth's surface, even down a deep mine or on a high mountain, and because few people are as yet familiar with newtons, most weighing apparatus is marked in grams and kilograms. So long as one remembers what happens on the moon, and understands why, the 'physics laboratory' difference between weight and mass is not important in everyday life.

Mass
Beam balances

Mass is measured by using a genuine balance, by balancing the object with known masses (grams,

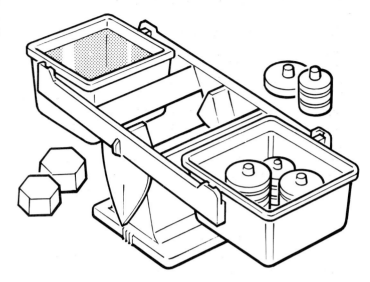

kilograms, etc.) unfortunately called 'weights'! The obvious kind of apparatus for this is a beam balance with two 'pans', 'tubs', 'buckets' (or even equal yoghurt pots) one at each end of a bar of some sort (even a well-balanced coathanger).

Young children may have their first experience of this sort on a seesaw, but it is only valid if the seats are fixed at equal distances from the central balance point (or fulcrum).

Beam balances have sliding adjusters, so that the central pointer can be set exactly to the zero marking; this should be checked before each use—for example, drops of water are not always obvious.

So-called moving fulcrum balances, like the sophisticated top-pan balances, can give excellent results if the mass is all that is wanted, but they are not good starting points for comprehension of the process of measuring it.

Standard masses ('weights')

Early experience is best gained by using large numbers of 1-gram plastic 'weights'. The concept of 'one gram' is important. The stackable red plastic type from E. J. Arnold is about the best, since they are a convenient size for handling, and can be built into firm sets of ten which can then be separated again to check the value. Some makers' plastic weights begin with 10-gram units, while Osmiroid Centicubes, first used as units of *volume*, are now offered as 1-gram weights for weighing (with probable confusion in the minds of young users).

On the other hand, after the early stages it is good to have a variety of 1-gram weights, e.g. brass, to establish the mass as the quantity. Balancing equal numbers against one another—brass against plastic—can be quite a surprise, but a valuable experience towards the difficult concept of density later.

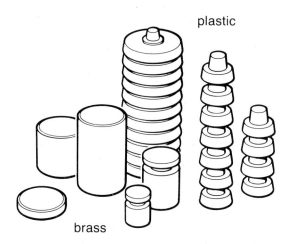

plastic

brass

Lever arm balance (Butchart type)

This direct reading balance is common in laboratories but needs to be completely understood and used with care. It works through a counterweight, not by balancing equal masses, but it is calibrated in grams, hence 'direct reading'. The main difficulty on first acquaintance is that there are two scales, one on the upper and the other on the lower edge of a quarter-circular strip of metal.

lever arm balance

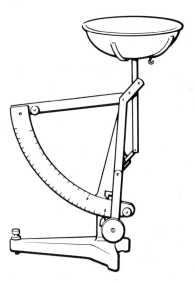

To weigh light objects, normally up to 250 grams, a solid circular counterweight is swung up on a flat metal strip under the pan. The feet of the whole balance are then adjusted so that the pointer is at the zero marking on the upper scale ('counterweight up, top scale') and the object is weighed to 1 gram accuracy. Heavier objects are weighed with the counterweight swung down and the lower scale used—up to 1 000 grams in 10 gram intervals. The balance has to be adjusted so that the pointer is at zero before each weighing, or at least checked to see that it does not need adjusting.

Straw or micro-balance

This is a Nuffield speciality, for weighing very small objects, or for detecting very small changes in weight. The units used as weights may be arbitrary, e.g. centimetre squares of paper, cut out of a sheet of squared paper specially for this purpose. It is quite easy to find the weight of these units, if necessary, by weighing some sheets and counting/ calculating the number of squares to the gram.

One end of a drinks straw is flattened and cut to a point, with the straw opened out behind the point to make a tiny 'platform' or scoop. The other end is loaded with a very small screw or scrap of Plasticine. The fulcrum consists of a darning needle pushed through the straw towards the weighted end; this needle rests on a small 'stand' made from thin metal cut to shape, or a wire frame bent to hold it free to swing like a seesaw.

Such a balance can have a vertical drawing-pin point to hold the paper weights at the one end, while the object is held in the scoop at the other. Alternatively, there can be a vertical scale beside the pointed end; this microbalance is an exercise in

straw balance

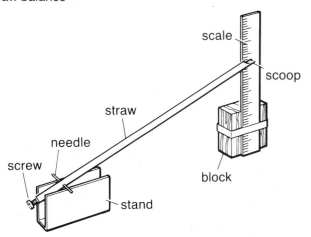

ingenuity. Uses are, for example, to demonstrate the loss of water by evaporation from a leaf or a small piece of damp fabric or blotting-paper under different conditions.

Bathroom scales

At the other end of the range of available 'weighing machines' is the bathroom scale—with a not normally visible mechanism of levers and springs. This is a direct reading machine, and needs adjustment before use, especially if the floor is uneven.

Weighing activities

There are many useful and important weighing activities in any science course; at the same time different pieces of apparatus can be used and compared, e.g. for accuracy and ease of reading, and different skills can be practised.

Here are a few suggestions:

Check for accurate weighing

(a) Adjust for zero start.
(b) Make sure the pan(s) is/are clean and dry.
(c) Test for sensitivity by adding small scraps of paper until the scale or balance reacts.

Very light objects

The weight of very light objects, e.g. sheets of air-mail paper, dried peas, single potato crisps, etc. can be found using an ordinary balance. If only one such object is weighed the inaccuracy may well be greater than the actual weight; if 10, or 50, or even better, 100 are weighed together, the inaccuracy will be shared out—and 1/100th of any inaccuracy may be completely negligible. Counting the peas is quick and easy if ten people count 10 each and pool them.

A measured volume of water

If plastic beam-balances are available, the water can just be poured from the measuring container into the pan. If water is weighed *in a container* it is only too easy to forget to subtract the weight of the container, or to forget to weigh it before it gets wet. . . Water, and damp materials such as fabrics, foodstuffs or soaked peas, etc. can best be weighed in a (watertight) plastic bag with a closure; a similar bag can be placed empty on the other pan, or can be hung on the spring scale beforehand—where it will probably not even register.

weighing water

Objects against beans

Books sometimes suggest beginning weighing by balancing objects against beans or other large seeds etc. Even young children soon realize that this is not the way adults weigh—and it is worth while to pick out say 20 of the smallest beans and 20 of the largest and compare the weights. This demonstrates the necessity for standard units of mass, and at the same time shows the variability of natural products.

Weighing and estimating

Since measurements in science are in grams and kilograms (and of course newtons), while everyday weights are still often in pounds and ounces, the 'feel' of 1 gram, 10 grams, 500 grams and 1 kilogram needs some practice. One cannot usefully estimate without considerable experience, so weights on packets and wrappers, and weights of well-known objects such as table-tennis balls, oranges (average size?), and drinks cans (full and empty) are good material. Any packets or objects which weigh just under (or exactly) 100 grams can be picked out—as objects which are being pulled towards the centre of the earth by gravity with a force of one newton!

Does the shape make any difference?

This is a 'fun' exercise with a real point to it: get a lump of Plasticine (rounded) and weigh it. Then make it into a large flat sheet, next into a very long thin worm, finally into many small 'pills'. At each stage, invite thoughtful 'guesses' as to whether the Plasticine will weigh more, or less, or the same as it did in the lump. Weigh it every time. Reason and visual impression may well be opposed, and not only for the very young. Try the same thing with long strings of 'Poppit' type beads, linked and then separated, or sticks of Centicubes. The Plasticine will probably have been the more misleading.

'natural' packaging

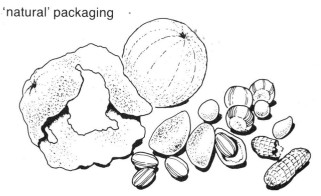

45

Making standard masses

Organize the production of standard masses from sand in small plastic bags with efficient closures. Some can be labelled, others left for estimation by 'feel'. Other objects can be assembled for 'less than 10 grams/more than 10 grams' and similar 'guesstimates'. Displays of objects all of one weight, e.g. a steel bolt, a bunch of keys, a large pile of expanded polystyrene, a bag of marbles, a clear plastic bottle containing coloured water, etc. lay good foundations for comprehension of the concept of density.

Using everyday materials

Make the most of everyday materials and interests in practising weighing skills and using various types of apparatus—e.g. 'How much of the weight is due to the skin/wrapping?' Try with oranges, bananas, eggs, Easter eggs (!). 'What wrappings are so light that they don't count?' 'What containers weigh more than the contents?' (conkers?, some glass bottles. . .).

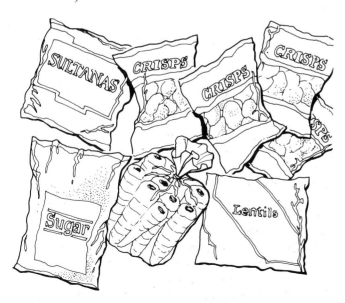

'How much water can a sponge/a towel/newspaper/sand soak up?'

'What weights do the bank scales test when they are used to weigh coins? What do the individual coins weigh? How much metal is worn off in (say) 20 years' wear?' . . . (N.B. if real silver is being weighed, the troy ounce is 31.1 grams, unlike the everyday ounce which is only 28.3 grams.)

'What are the important weights for letters and postage? Inland, abroad, airmail? What kinds of scales do they use at the post-office for which kinds of objects? When does a letter turn into a parcel?'

Looking-up activities

These can include luggage limits for air travel, etc. Jockeys and boxers have to be very careful about their weights. What are the limits? Where do colleagues and family members fit in?

What are the weights of interesting living creatures? (Largest mammal, the blue whale—150 000kg; largest land mammal, African elephant—4 to 6 tonnes; pygmy shrew—1.6 gram; kangaroo baby when born—less than 1 gram; humming-bird egg—0.25g; ostrich egg—1600 g and how many chicken eggs would that be?).

Weighing air – a special problem

Of course air weighs something—the miles of atmosphere above us make a pressure equal to 1 kilogram force on every square centimetre—but actually weighing a little air is not easy because it has so little mass. Books often describe 'An experiment to show that air has weight' or even 'To weigh some air'. These usually begin with two balloons on a balanced beam, one of which is then blown up, or two inflated balloons on a beam, one of which is then deflated. In neither case is the weighing valid, since the inflated balloon displaces almost as much air as it contains, and this reduces its apparent weight almost to that of the rubber alone, i.e. to that of the uninflated balloon.

Blowing up the balloon by mouth may make it slightly heavier—but it is found after the event to be wet inside.

Much more air has to be compressed into the container to get a true demonstration that 'air has weight'. This can be done using a strong sealed can with a valve in the screw cap. It is also said to work with a car tyre inner tube, if such can be found and pumped up.

'Weightlessness'

Our weight, and the weights of other things, can be felt. Weight is the result of the pull of gravity, pulling us and all the other things towards the centre of the earth. So whether we are standing on the floor, or the street, or the floor of a lift, our weight is there all the time. If the lift goes down rather fast, we feel as if the bottom had fallen out for the moment. When this happens, the floor of the lift is going down just a bit faster than we are, so it isn't holding us up. Our feet are not pressing on the lift-floor at that moment, so we feel as if we don't weigh anything. This is the feeling of weightlessness.

But the force of gravity is still pulling us downwards as hard as before, so we will have exactly as much weight as we had when the lift was at the top landing, and as we shall have when it gets to the bottom.

There is no such thing as 'weightlessness'. It only feels like it. The correct term is 'apparent weightlessness'.

In a rocket or space-ship what looks like weightlessness lasts much longer. The people inside seem to 'float' around, and so do their meals and the tiny ends of hair when they shave. That's why they often eat out of tubes, and shave with electric shavers with tiny built-in vacuum cleaners. But the effect of gravity is still there, though not as strong as on earth. The pull downwards is being neutralized by the movement of the rocket—in fact, it is just the 'lift-going-down' effect, lasting much longer.

This can last for any length of time if the space-ship is programmed to be a satellite of the earth, going round and round at the same speed—just fast enough to prevent it from being pulled back on to the earth (by gravity, of course).

The fact that the 'lift-going-down' feeling only happens at the beginning of the descent gives us the hint that it is due to the downward *acceleration* of the lift. The same kind of sensation, again due to acceleration, can be felt if one is standing in a bus or Underground train as it starts again after a stop.

When the lift, or the train or bus, is stopping, it decelerates and produces the opposite sensation, i.e. in the lift we feel as if we weigh more than usual. We keep going while the lift is stopping, and our feet press more 'heavily' on the floor than usual. If we are standing in a train or a bus and leaning against the front end of the coach, we find ourselves being pressed against by the coach end, because it is stopping before we are. Very interesting and convincing tests of these phenomena can easily be carried out, especially in the Underground. All one needs is a clear plastic (or glass) tube, or a slim bottle with parallel sides (no ribs). Insert a marble. Close the tube, place it lengthways on the 'window-ledge' and hold it still while the train starts and stops at several stations. Bus and car travel can provide equally good tests, but they do not often have the frequent starts and stops which show the results most clearly.

Apparent weightlessness can be shown in a lift which drops rapidly at the beginning of its descent. An object hanging freely on a spring scale seems to lose at least part of its weight as the lift accelerates, and the pointer rises on the scale.

The dependence of apparent weightlessness on acceleration of the vehicle is more difficult to understand in a space-ship circling the earth at a steady speed. It can be calculated by a slightly sticky bit of mathematics that the circling capsule is accelerating towards the earth all the time, otherwise it would head out into space. This does not mean that it is heading towards the earth, and a collision; the acceleration (due to gravity, of course) just keeps it in orbit.

There is no point between the earth and the moon, or indeed further out, where gravity does not act.

The force of gravity of the relatively small moon is strong enough to haul millions of tonnes of sea water up on our beaches twice a day, for example. The moon makes our tides.

Mass and inertia

The mass of an object gives it the property called inertia—the reluctance to accelerate when a force is applied to it. A heavy object near the top of a slope may not start to slide down, even though once it is sliding it will go on sliding to the bottom (i.e., it is not held back by friction).

This can be tested very easily.

The wheel of an inverted bicycle needs a definite push to get it to spin, though once it is spinning it may go on for a long time.

A car with gear in neutral, engine switched off, brakes off—on a flat road or car-park—needs a considerable force to get it moving, though once started it may keep rolling. All of these are examples of the effect of inertia, which depends on the mass of the object.

Car wheels which are not 'running true', i.e. spinning evenly, have small masses of lead (with the weight in grams marked on the side) clipped on the rim. These balancers add their inertia where needed to make the wheel turn evenly.

Inertia is clearly linked to mass, not weight, because it acts in all directions, including horizontally. It can be experienced on a small scale by rolling balls of different materials but equal sizes, e.g. a large glass marble, a plastic 'bouncy' ball, and a table-tennis ball along a horizontal gully (a length of plastic track for a sliding door is good) giving each the same standardized 'push', e.g. a flick with the thumbnail.

Stopping moving objects also involves their inertia—the greater the inertia on the move, the greater the reluctance to stop. This is a different problem from starting the movement, because the velocity is also involved. A large mass (with a large inertia) may be easy to stop if it is moving very slowly—its momentum is equal to its mass multiplied by its velocity, so both factors affect the force needed.

Time spent observing by a railway shunting yard provides examples.

Time

Time as such cannot be measured. What can be measured are events, mainly movements, which are regular enough to indicate 'how time passes'.

The modern name for a piece of apparatus which changes a 'signal' in one physical form into a corresponding signal in another physical form is a transducer; transducers are very common in time-measurement, e.g. in the cuckoo-clock, the swings of a pendulum are changed into movements of the hands on the face, and the occasional bird-call.

Available equipment

Sun and shadow 'clocks'

These are good within their limits, provided that they are accurately made and correctly set up—and that the sun shines.

Water and sand clocks

These, like sun clocks, are of great antiquity, and are of limited use. The water clock depends on either water dripping out of a container through a very small hole, or on water entering a floating container. The water-drip clock goes 'fast' at the beginning, and 'slow' at the end, when the water gets low and the pressure is therefore less.

This is worth testing.

The sand clock is either a commercial 'timer', e.g. for boiling eggs, or an instructive but seldom useful (amateur) model, which tends to stop dead because of a slightly oversized sand grain.

Candle clocks

These have a long history, though royal patronage was probably withdrawn some 1100 years ago. They make excellent material for testing scientific variables, e.g. the type of wax, the presence of draught, and the diameter of the candle. For quick and satisfying tests against a clock, small birthday-cake candles or halves of tapers, with a firm non-inflammable base, give good results. Several forms of marking are available, either visual (wax crayon), or audible (map pins, falling when their time comes on to a tin-lid).

Pulse for timing

For many years the human pulse was used for measuring time—and was considered reasonably accurate. Certainly it is better than guessing—a minute 'feels' very different under different conditions.

But the human pulse-rate varies with age, with exercise or excitement, with infection, etc. so it is rather to be timed against a clock rather than a source of timing itself. This timing of the pulse-rate is, however, important. Many people do not find their pulse in the wrist easily, so if a class is about to do the exercise it is much more satisfactory to let each begin counting when ready, looking at a wall-clock with a seconds sweep-hand and waiting for the vertical position, rather than to expect everyone to begin at the word 'Go'.

The same is true for the timing of breathing rate, though here we find much wider variables. Some individuals breathe more slowly (but more deeply) after vigorous exercise, some can hold their breath for a whole minute, especially if challenged. The consciousness of 'being timed' affects some people's breathing rate as well. It is useless as a timing mechanism, but important as a technique—remembering to remind the subject(s) that 'in and out' is only one breath.

Pendulums

In Galileo's time a 'doctor' carried a pocket pendulum—a 'pulsilogium'—for timing the patient's pulse. A pendulum is still used as a basic method of measuring time accurately.

The simplest pendulum, a 2-metre length of cord and a lump of Plasticine (safer than the metal pendulum-bob, and more useful for investigating variables) can be hung freely from a hook in a door-frame, after checking that the door can still be opened and closed. Hypotheses can be made and tested on several possible variables, e.g.

(a) the mass of Plasticine in the 'bob';
(b) the extent of swing—above horizontal, horizontal, or with only a small displacement before letting it swing;
(c) the shape of the Plasticine bob—spherical, flattened; fitted with 'fins' or plastic bag streamers, i.e. with increased air-resistance;
(d) the length of cord from the hook to centre of the bob.

The variables (c) and (d) are the ones which will make the difference; (c) is easily explained, leaving (d) as the fundamental factor. This is then available for testing—halve the length of cord, shorten it to one-quarter, etc. In every case, time a set number of swings—there and back—and repeat. A 1-metre length of pendulum should give a 1-second swing in each direction, i.e. a 2-second complete oscillation.

If at all possible, a metronome with an open front should be studied. It is a more complex form of pendulum, loaded above as well as below the turning point of the pendulum—and of course is marked with various times of swing.

metronome

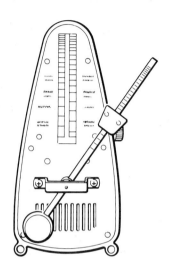

Pendulum clocks

The grandfather clock shows how the time of swing can be adjusted—a very small upward movement of the large pendulum bob by means of a nut screwed up on the rod makes the clock go a little faster, a slight lowering makes it run slightly slower. Expensive mantelpiece clocks in glass cases may have a vertical rod turning to and fro, with brass spheres attached whose inertia keeps the rotations accurately timed. The mechanism looks rather like the governor on a steam-engine.

'Ordinary' clocks and watches

These have a balance wheel which spins in each direction, driven one way by the wound-up spring and back by the very fine hairspring. This mechanism is worthy of close observation, and perhaps timing either from the face of the clock/watch itself, or an external reference.

The dial clock with a sweep second-hand is a most valuable timing instrument, and about the only way in which one can credibly see 'the passage of time'. It can be used in many situations where a stop-clock or stop-watch would also be appropriate, though of course not for sports events.

Stop-clocks and watches

The advantage of the stop-watch is that it records the time taken for an event; its disadvantage is the time needed for the timekeeper to react and press the button. This is said to be about 0.3 second for experienced timekeepers; for the inexperienced there may be no recording, owing to the wrong button having been pressed. Practice is essential, with frequent checking, especially if there is any likelihood of a colour-blind person doing the timing, as the buttons are usually green and red. For static events, e.g experiments, or for sports events which start and end at the same point, the button-pressing delay should cancel out; both start and finish could be 0.3 second late. . .

The same kind of practice is necessary if the laboratory stop-clock, the so-called 'tombstone' stop-clock (because of its shape), is to be used. The lever is either up or down—check which—and the

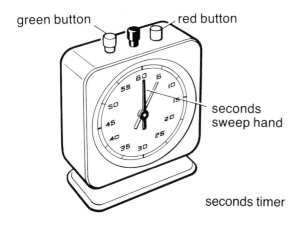

green button   red button

seconds sweep hand

seconds timer

'return-to-zero' button is left severely alone until written records have been made if required.

Digital clocks and watches

These have considerable advantages over the dial types; the numerals are there to be read, and no 'translation' is required. However they make two main assumptions:

(a) that the user is familiar with the consecutive numbering of the minutes all round the dial, e.g. 8.49 rather than 11 minutes to 9; and
(b) that the 24-hour clock will be used.

For many people the second of these demands some effort—subtracting 12 from the hour value when checking. Practice on a dial clock with the second ring of numerals (as on many railway stations) is a help; so also is study of all kinds of timetables. Quick calculation of time intervals is often necessary, and this can become easier with practice, e.g. the intervals between 13.01 and 01.13, between 19.05 and 05.19, between 21.22 and 22.21, and between 23.59 and 00.01. In the early stages these may really have to be thought out.

Finally, since timing is such a demanding skill, the more events which can suitably be timed for experience the better: e.g. the rate of sweep of windscreen-wipers, often by the squeak-scrape sounds (remembering that some cars have two speeds available); the timing of the telephone ringing tone (brr-brr, brr-brr. . .); the exact length of time paid for by an advertiser for a television advertisement (the advertiser would know!); the time taken for the 'clapper' sound made by two pieces of wood to travel the length of the playground to a high wall and back again (i.e., timing an echo, and then finding the speed of sound).

One need never be short of practice material.

Angles

Measurement of angles is exceptional, because the common unit is not a decimal quantity. An SI unit exists—the radian—but it cannot be measured with an ordinary protractor, since 1 radian is approximately 57.3 degrees. 180° = Π radians.

However, we manage very well with 90° to the right angle, 180° to the straight line, and 360° for the whole way round a point, back to the starting position, i.e. 'once round the clock', which is a rotation.

'Square corners'

One angle is immediately recognisable, and measurable, even without a protractor. This is the right angle. Everyday life is so full of 'square corners' that even young children know them at sight, and most people are uncomfortable when a corner which should be a right angle is just not right.

Testing a right angle only needs a piece of paper; and it looks most impressive—almost like magic—if an irregularly torn oddment is folded in approximately halves and then in quarters, producing a perfect right angle so long as the folding is done carefully, 90°, exactly.

Fairly thin paper can be folded again to halve the angle, though 45° is not nearly as easy to judge by eye as 90°. Further folding is quite unreliable. The folded paper, unfolded again, demonstrates that 4 right angles make up one 'full turn', all the way round, therefore 360°.

Protractors

The common semicircular protractor has drawbacks

(a) There is usually a strip along the straight side which has to be ignored, since, like the unmarked bit on the end of the ordinary ruler, it is only there to protect the line which matters. This means, however, that the base line of the angle to be measured has to be viewed through the plastic, and the actual 'point' of the angle has to be very carefully fitted to the centre of the measurement part of the semicircle.

(b) Obviously a semicircular protractor can only go up to 180°, so that even if the protractor is inverted to complete a circle, the numerals for the number of degrees will start at 0° twice instead of giving the true value for anything beyond 180°. Many British protractors give the values (as far as they go) from both sides of the centre; some Continental protractors only count from the right.

Circular protractors

However, protractor makers are now producing circular instruments, marked all the way round from 0° to 360° in both directions, with diameters to place over the base line of the angle to be measured. These circular protractors are a great help in both measurement and comprehension of angles (to 180°) and rotations (beyond 180°).

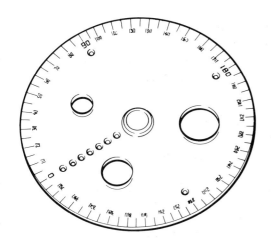

'Angle measurer'

A special version of the circular protractor is the 'Angle Measurer', with sliding discs and moving arrows, so that the angle to be measured is actively 'moved through' to show the amount of rotation involved. This makes the concept of an angle clear in a way that the static protractor cannot do.

All three types of protractor should be tried, but for learning purposes probably in the reverse order of the above list, with the traditional semicircle last.

'Pie charts'

The so-called pie chart is a favourite tool of statisticians and advertisers. It is very easy to grasp its meaning in a general way, but the arithmetic involved in converting 360° into percentages, or percentages into fractions of 360°, is tough for most people. Luckily for today's students, there is also now available a circular Pie Chart Scale, similar to a protractor but marked in percentages from 0% to 100% clockwise round the scale.

Angles, rotation and 'turns'

Some mathematicians regard any measurement beyond 180° as not being an angle any longer, but a rotation. Others call anything greater than 180° but less than 360° a reflex angle. One accepts both versions, and measures with a circular protractor. In everyday language *turn* tends to mean 'at a right angle' unless an angle is named; '*turn round*' tends to mean 180°, i.e. facing in the opposite direction, while '*turn right round*' can mean 360°, i.e. a complete rotation. 'Turn right round twice' would therefore mean a 720° rotation.

Examples of rotation

Simple technology, like biology, is full of examples of rotation. Try the screw top of a coffee jar for a specimen measurement:

(a) Turn the loose lid 'backwards' (anticlockwise) on the jar until it just clicks and drops on to the thread.

(b) Hold it, and mark jar and lid at the same point, e.g. with Tipp-Ex.

(c) Screw the lid on (clockwise) until it is as tight as it should be, counting the rotations and the final part. How many degrees in total?

(d) Holding the jar down, unscrew the lid until it just comes free—to check the first measurement.

(e) Look at the glass thread on the neck of the jar, and trace it round the jar—how many turns of the jar neck? This is almost certainly a single-start thread, a continuous helix. Look (or feel) inside the lid. Look for other examples; most people have one lid which seems to need several turns, often on an older jar.

(f) Find a recent jam or marmalade pot of the screw-type, but probably with inward projections in the lid rather than a screw-thread. Mark and test this for rotation before it is tight. It may only need a turning angle of a few degrees—measure this angle, and again check.

Look at the glass thread on the jar; it is likely to be a set of short multi-start threads, each only a short distance round the jar. Look for other examples; consider the advantages and disadvantages of each.

multi-start thread

Angles and clockfaces

Clock dials offer a variety of angles, though the numerals are not directly related to the angular measurement. Beginning with the first right angle, opposite the hour 3, the relationship is easily calculated; from 12 (which is also zero in terms of hours, 00.00 hours) round to 3 is 90°, so a 1 hour movement of the hour hand goes through 30°. This is in fact one of the few everyday places where one can see a guaranteed angle of 30°, which is not an easy one to estimate visually. With a clockface model the hour hand can be moved round 30° at a time, keeping the minute hand vertical to indicate the beginning of the angle moved through. Check, at the end of a complete rotation—12 × 30 gives the expected 360°.

A full-sized outline of the face of 'Big Ben', 7 metres across, on the playground, gives opportunities for measuring out the appropriate angles. One diameter of the 3.5 metre radius circle acts as a starting line: a right angle can be drawn from this using a large sheet of newspaper for a convenient source of a square corner. Protractors are useless on this scale; it is obvious that an error of only a degree or two at the centre will be seen to be wildly out at the edge of a circle of this size. But angles of 60° (for 2 hours) are needed; perhaps the best method is to step them out around the circumference with the 3.5 metre long string which was used in drawing the original circle. This is a familiar technique to anyone who has drawn '6-petalled flowers' with a pair of compasses. Each 'clock minute' will, of course, represent 6°, but this is too small an angle to draw on a coarse model of this sort; the 60° angles will not be absolutely accurate drawn this way, but will be near enough.

Compass directions

To avoid ambiguity it helps to specify a magnetic compass as distinct from a pair of compasses.

Out of doors, preferably in the middle of the playground, away from iron drainpipes and manhole covers, the cardinal points of the compass can be established. Again, as with the protractor, if the lines defining the directions are too short, a small inaccuracy will make a considerable error as the lines are extended. One way to overcome this is to have two magnetic compasses and a long straight rod—perhaps a 2-metre length of dowel rod. If the rod is lined up with one compass at each end, say pointing north and south for a start, then the direction is likely to be reliable. A right angle from this line can be drawn by using the sheet of newspaper, and checked with the compasses. By folding the newspaper right angle in exact halves, a 45° angle can be added, and repeated round the centre point. Amateur (or professional) mariners will want to add smaller angles, but probably by measuring equal distances from adjacent lines already drawn—checking, of course, with a compass at the centre.

The orienteering compass

This, whether Silva, or a similar bearing compass, combines magnetic compass and protractor. Many activities can be triggered off if such apparatus is available, e.g. true orienteering, 'Turtle'-like movements with a human 'Turtle-actor' following computer-type instructions, and much map-work.

Good starting questions, provided suitable maps are available, would be 'What direction would a/ the flying crow take from Land's End to London?'

'And from London to Land's End?' 'Which way from the Isle of Man to the Isle of Wight?' and then 'How many degrees west of north from London to Glasgow?' 'And from Birmingham to Edinburgh?' and so on. A whole series could be constructed for practice, e.g. Aberdeen to Aberystwyth? Birmingham to Brighton? . . . each establishing the skill.

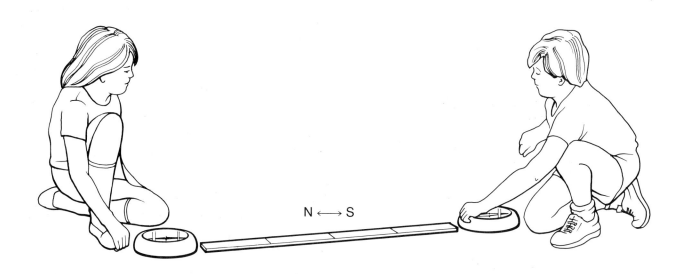

N ⟵⟶ S

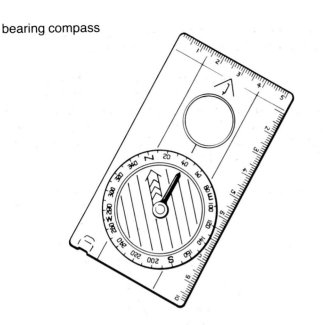

bearing compass

Horizontal, vertical and perpendicular

Many people find it easy to confuse two, or all three, of these terms, due to early confusion when all three were taught at the same time.

Horizontal

This is a good starting point. Because of the different stress on the two words, it is not always obvious that the horizon is the reference.

Illustrations in travel brochures offer numerous examples of dark blue sea meeting lighter blue sky at a straight line in the distance—here it is. Of course, we know that the earth is round, but it does not show, or matter, in this context.

Useful vocabulary can arise. If the answers to these questions are drawn on a map, each represents a vector quantity, i.e. a displacement (as the crow flies) in a stated direction. This can be plotted on the map, with direction and distance.

There are many ways to test a horizontal surface, e.g. a table-top or pavement. A marble not only rolls of its own accord, but demonstrates the direction of the slope if any; water similarly shows, say in the playground, the way down—perhaps to

horizon between sky and sea

the drain. The professional's spirit-level is an elegant tool for the purpose, and can be used where primitive methods are not suitable.

Vertical

The vertical direction is simply the direction of the pull of gravity towards the centre of the earth. A small but relatively heavy 'plumb-bob' on the end of a flexible line, cord or thread, demonstrates it, and the relation between horizontal and vertical directions can be checked if the plumb-line is held beside but just not touching a horizontal table-top. Right angles of this kind can be tested with a set square, a T-square or a try-square.

Perpendicular

Two directions perpendicular to one another are, of course, at right angles to one another, but can be on a horizontal sheet of paper, or up in the air or on any plane even if a sloping surface.

Gradient

Gradient or slope is a quantity involving an important angle measurement—both on the large scale for roads, contours, etc., and on a small scale for graphs.

There are, however, different definitions of the term. It can mean the actual angle between the sloping line and the horizontal—the slope angle—or it can be a calculated ratio between the vertical rise and the horizontal distance or the sloping distance. Mathematicians generally use the second of these, the geographers' and roadway version being the third. If traffic permits, it is interesting to try to find the slope, or gradient, of any available ramps, e.g. for prams, delivery trollies, wheelchairs, and for cars into garages. Measurement is easy if the side of the ramp is available, and can be checked by drawing a scale

diagram, or by cutting a 'paper pattern' and measuring the angle with a protractor.

The angle obtained can be checked with the instrument made for the purpose, a clinometer, whose base is placed on the sloping surface, and the freely-swinging 'plumb-line'/rod points to the angle on a graduated scale like a protractor, with 0° in the centre of the curve.

clinometer

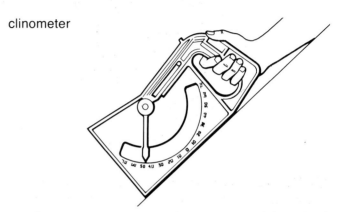

Card gauges

Many other angles can be estimated, though not measured, e.g. the slopes of roofs, car-ports and greenhouses. One possible way to compare such slopes is to construct and cut out card 'gauges', i.e. angles, which can be held up and viewed against the end-on aspect of a roof; it is possible to select the best fit in this way, and to compare all visible gable-ends and lean-to roofing (a) to see the differences—the steepest, the flattest—and (b) to find the approximate angles.

Card gauges are useful also for testing angles where it is not possible to get a protractor into the right position. Here are just two examples: a filter paper when folded flat makes a 90° angle, but this is not the angle of the funnel in a section diagram. Cut a card gauge to fit the funnel. Some limpet shells are flatter than others; cut card gauges to test them!

card angle-gauges

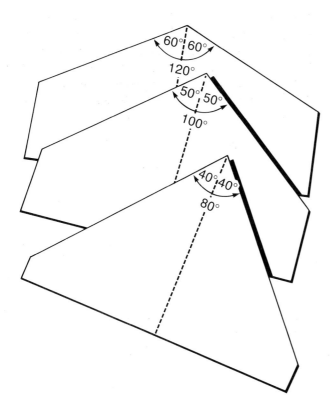

Reflection and refraction

The science of reflection and refraction of light depends on accurate measurement of angles. The most important of these is once more the right angle—a perpendicular drawn to the surface of a mirror, or a block of glass, or a prism, or to a tangent to a curved surface such as that of a lens. In each case, the perpendicular drawn (in the diagram, not in real life, of course) is called the *normal* at that particular point to that particular surface.

The rest is easy, with a protractor. The angles which have to be measured are the angles between the rays of light and the *normal*, not between the rays of light and the surface of the mirror or block of glass.

Reflection

A coarse model shows how a ball bounces back from a hard smooth surface—rays of light 'bounce back' in exactly the same way. If the ball is rolled (just hard enough) along the floor or smooth tabletop, to hit the flat wall behind it, one knows from everyday experience where to expect the ball to bounce back (or one can find out, if the experience has not previously been consciously registered).

Quite young children have sometimes discovered that light (usually sunlight) can be reflected from a small mirror or a bright tin-lid in the same sort of way (maybe into the teacher's eyes. . .). By directing a narrow beam of light from a torch or special ray-box through a narrow slit in card, the track of the beam can be observed and drawn—to a mirror, and away again as it is reflected. The normal is drawn on the paper, at right angles to the position of the back of the mirror (where the reflecting surface is) if it's a glass mirror, front if it's plastic, and exactly where the narrow beam of light hit it. The angles needed are those between

beam of light
going into water

the light beam and the normal; these angles should be equal, and on opposite sides of the normal—one going to the mirror, and the other coming away from it.

Exactly the same pattern is found when blocks of glass, including prisms, reflect light; this can happen on the outside of the glass, or inside the glass when the light hits the far side.

The surface of water often acts like a mirror, in that reflections are seen from the upper surface. Careful experiment demonstrates that the under-side of the surface of water also reflects light downwards—so that fish only see out of the water through a circle above their heads—beyond this they get a reflection of the waterweeds and mud around them.

The kaleidoscope

One of the most satisfying uses of reflection involving angles is the kaleidoscope. Beginning with two mirrors facing inwards at an angle of 90° it is clear that the object (small toy, beads, etc.) between the mirrors, together with the images formed by reflection, fill a circle. This would suggest that there are three images and the real object, filling the 4 right angles. This is not logical, since each mirror is producing an image of the object, and an image of the first image as formed in the opposite mirror. This would result in 4 images—which is what actually happens; the two 'second reflections', the images of images, come on top of one another 'at the back', so that they look like one image filling the fourth quarter of the circle seen.

And what happens if the angle between the mirrors is 60°, which is a common angle for commercial kaleidoscopes? The circle appears full to the viewer, giving 60° for the real object, and apparently 5 images (and images of images). Again,

the last pair of images, one from each mirror, just coincide 'at the back'.

Bringing the mirrors closer together, i.e. making the angle between them smaller and again smaller, has predictable results once the first two angles have been tested. It is a pleasing exercise to draw other measured angles, to predict how many images will be seen in the two mirrors, and to test one's hypothesis.

principle of the kaleidoscope

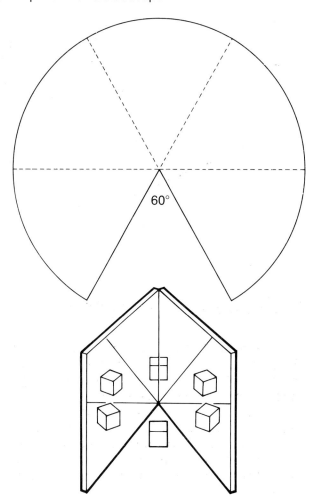

The optical lever

An important use of reflection is the 'optical lever'. A spot of light is directed on to a very small mirror attached to something which is going to turn slightly, e.g. the coil of a galvanometer. The mirror sends back the reflected beam, and this turns through twice the angle of the mirror movement; test it!

simple optical lever

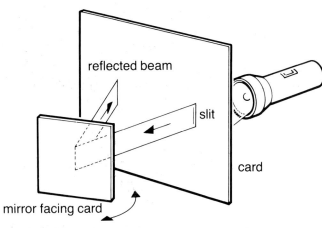

Curved surfaces

images in convex and concave mirrors

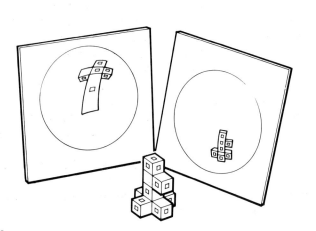

When light meets a curved surface such as the bowl of a soup-spoon or a curved mirror, exactly the same thing happens as when it meets a flat (plane) mirror—the light is reflected, making the same angle with the normal as the angle at which it arrived, but on the other side of the normal. The normal in this case has to be the perpendicular to the (straight) tangent to the curve at the point where the ray hits it. 'Peculiar' images are seen, but the light is following the very simple rule.

Refraction

Refraction is the bending of light as it goes from one transparent substance (say air) into another (say water). This can be seen when a narrow beam of light from a strong torch (or a projector) is shone into water containing a very little red ink (eosin) or a few drops of homogenised milk, in a partly darkened area. A small fish-tank is a good container, since the flat sides do not interfere with the result as the curved sides of a jar would do.

The two angles that are measured—light going towards the surface, light continuing in the water after bending at the surface—must once more be

beam of light going through a block of glass (top view)

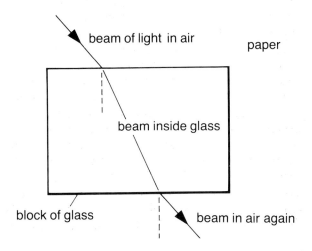

those between the beam of light and the normal (the perpendicular to the surface). A protractor held against the side of the tank can give quite a good result. The generalization one hopes to get out of several results is that the light bends towards the normal as it goes into the denser substance (i.e. water).

More accurate measurements are made using a rectangular block of glass. If this is lying on a sheet of paper, the direction of the beam inside the glass can be seen at the bottom, and here the beam can also be traced as it comes out of the denser substance, where it bends away from the normal as it goes back into air. As with reflection, the actual values of the angles will depend on the angle at which the beam of light arrives at the surface. This, of course, can be planned and measured in advance, and the whole diagram can be plotted on the underlying sheet of paper.

Sun and shadow-dials

This topic provides many interesting angles—to be discovered as well as measured. Whether the base of a sun-dial is horizontal, with the triangular gnomon standing up with its front point to the centre, or it is a so-called equatorial sun-dial with a sloping base and a rod casting the shadow, one angle of slope—at the front of the foot of the gnomon or at the front edge of the sloping base—has to be fixed at a definite angle. This angle depends on where the sun-dial is to be used. It is the same numerically as the latitude of the place: for example, at Land's End the angle is 50°, in London and Cardiff 51° 30′ (51½°), at Derry and Newcastle on Tyne 55°, and at Inverness 57°30′. In the West Indies the angle would be 10° for Trinidad and 15°30′ for Jamaica.

Then there are the angles marked out on the base-plate, whether it is sloping or horizontal, to show the time in hours and fractions at which the shadow

falls. The hour lines are spaced at 15° apart, but one would have to add a temporary scale to deal with Summer Time.

The famous bronze sun-dial (*above*) was made by the sculptor Henry Moore in 1967 It is in the City of London and belongs to *The Times* newspaper. It looks like a very curved bow, with the bowstring casting the shadow on a semicircular strip scale. There are many other alternative forms, and anyone with the interest could invent a new variety—starting from a vertical pole in the playground, and discovering the angles by observation. An early activity would be to make a large protractor—as noted before, the small desk type cannot be used accurately on angles between long 'outdoor' lines. On a large sun-dial, the time can be read to five minutes if the marking has been done accurately; a project to make one could be a year-long activity.

Living things

Special characteristics

Measurements of people, plants and animals have special characteristics which are different from those of other sets of objects.

Continuous variation

Measurements of living things usually show continuous variation between limits. For example, when investigating human heights there are none below 10 cm, and none above 500 cm. But in a primary classroom there are closely crowded values somewhere between 80–150 cm, with no neat 'steps' of the kind one gets in manufactured products. The individual being measured can also vary on the spot—try asking 'Can you make yourself a tiny bit taller?'. . . . The general rules for scientific measurement—measure more than once, and check every measurement—are more obviously needed here than in most other fields.

Many sets of measurements of living things are the raw materials of simple statistics. It is very easy to develop the ideas of the 'average' (in statistics the **mean**), the middle size (or **median**), and the most 'popular' (or **mode**). Children will enjoy finding these, and there is no problem of the child who is unhappy to be the very largest, or the very smallest.

One can also show with convincing (chosen) sets of results that statistics can only be at all reliable if the number of 'specimens' measured is large, and chosen as nearly randomly as possible. By 'picking all the big ones out'—whether children or crisps—it is very easy to demonstrate 'lying with statistics', e.g. 'the average of ten'—yes, true, but which ten?

Growth

Living things have other characteristics which are uncommon elsewhere, such as growth. Here there are most interesting ways of looking at measuring activities. There is the straightforward week-by-week height measurement of a sunflower stem—or better, ten in a row, getting equal sunshine and watering, etc. Or better still, ten here, and another ten in a shaded place. Naturally all the seeds should come from the same 'head', from last year's plants.

Then there are measurements of heights, perhaps registered on the doorframe as in some country kitchens, of a growing member of the family, often the first child. . . And valuable evidence of 'growth' in a general sense can be observed from the mean, median or mode (or all three) of heights taken from the year-groups in a school. Individual children can see for themselves the height that they may expect to reach when they are as old as father, or mother, though here they will notice—have probably already noticed—the common difference between the heights of the two parents. Some few may have a mother taller than father, in which case they come back to the idea of 'large numbers needed for statistical results', i.e. fathers on the average are taller than mothers on the average. If photographs of families with several children can be found, the 'steps' in height are often beautifully demonstrated, especially if the births of the children are fairly evenly spaced.

Other factors about growth measurements of living things may, and should, be discovered and comprehended—e.g. that growth does not go on 'for ever'. For most kinds of living organism there seems to be a maximum, however much older the organism may become. Some individuals may never reach this and some may add weight long after they reach their full height.

Optimum size

There may also be a different consideration—an *optimum* height (or weight, etc. of course). The idea of an optimum size is a most important one in other fields, from cars to calculators. The question here is 'optimum for what?'—and the answers demand a good deal of lively lateral thinking. One could suggest that the optimum size of car is not a Mini for a 6 foot 6 inch man—the most exaggeratedly Z-shaped driver. On the other hand, for a small person needing to park in town streets, what could be more appropriate?

Recording measurements

Measurements of variable material or of growing organisms are well displayed by simple graphical means. Actual heights or lengths can be shown by paper strips or chalk marks on wall-boards; these can be turned into book or file records in the form of block graphs—and all of these are more convincing than lines or lists of figures. Such graphical demonstrations of results help in a quite marked way towards the discovery of exceptions; the result which does not 'fit' may well be due to inaccurate measuring—not so easily spotted among other written figures. The rate of change is also much easier to see, for example in the growth of sunflowers or hollyhocks, which slows down to zero as the flower heads or flowers expand. The height of the sunflower actually decreases towards the end, as the flower-head hangs over.

If the record of sunflower height is kept as points on graph paper, joining up the points gives a visual reminder of the actual events.

Many people are attracted to the 'record' idea—the tallest, the longest, the heaviest, the oldest, etc. Lively interest may be stimulated by using the *Guinness Book of Records* (but some other Guinness books may not be so reliable for 'scientists'). It should be clear, however, that the record-holders for some items, such as the heaviest man or woman, may also be extremely unhealthy—and to be pitied rather than emulated!

People-measurement

Personal measurements have some great advantages, in spite of the difficulty of getting them accurate in mathematical terms.

First, the material is available—and often in quantity. Second, the measurements are interesting to the owner. Hence, thirdly, they are likely to be carefully observed and checked. In addition, if only a few measurements are suggested as 'starters', the activity gives plenty of opportunity for enterprise and ingenuity.

What can we/they think of to measure?

Height

This is the obvious one, but there can be some small point to make it a bit different, e.g. comparison with the marked-up height of a very tall well-known personality—as a kind of challenge. The usual precautions should be taken, of course; no heels (or better, no shoes), and a true right angle (book, set square or the official sliding marker) on top of the head. A tape-measure can be stuck up the wall with Blu-Tack, with the bottom end starting exactly at floor-level. Probably two tape-measures will be needed, and a little addition, unless a long one can be found.

Feet

Foot length, without shoes, can well be followed by incidental testing of guesses (hypotheses) such as 'Do the tallest people have the longest feet?' and 'Are middle toes, like middle fingers, the longest?' Shoe sizes are worth a short investigation, and are not too simple—partly because of the variable extra length of welts, etc., and partly because of the change-over from 'children's sizes' to adult sizes. Do we know any adults with small enough feet to wear children's sizes? And any members of the class who need grown-up sizes? It is worthwhile to borrow a shop shoe-gauge for testing the correct lengths, though it may not be worth buying one.

Foot widths are variable too, and change when the owner stands up (putting more weight on the feet). This needs to be considered when new shoes are bought. Some shops have quite a range of widths,

but perhaps in the more expensive makes. Even the popular lace-up canvas and rubber sports shoes may not be wide enough for some people. Good simple statistics can be extracted from foot lengths, e.g. a comparative study of all the 8-year-olds, or the 13-year-olds, or the boys' measurements compared with those of the girls in exactly the same age group. Growth also shows up well, especially if the foot length of a new baby can be compared with that of its parent of the same sex; a superimposed outline makes the point visually.

And what about the origin of the unit of length—how many people can a class discover whose feet would give an accurate measurement, heel to toe, across the school hall? Or perhaps more conveniently across their own bedroom floor? This will be a question of tall fathers 'putting their best foot forward' . . . the distance being checked by offspring with tape-measure.

Hands

Hands are very useful for rough measuring—for tables, boxes, shelves and spaces to fit them into. A hand-span along a ruler is easy for all but the youngest, who sometimes have not mastered the muscular action of spreading fingers, and have to do it with the other hand. Again, like other personal measurements, hand-span can be extended with effort, and aspiring pianists will want to check their span against an octave on the keyboard. Young violinists and guitarists, left hand only, may like to measure the size (diameter or circumference) of the ring they can make between tips of thumb and middle finger. This is much the same as the size of tip-to-tip grasp, which can be tested round a largeish lump of soft Plasticine. The O-level Nuffield texts suggest measuring the length of your middle finger: put your hand down like a tent, and measure from the middle knuckle—the top of the tent—down the slope. Grasp-quantity, either by volume or weight

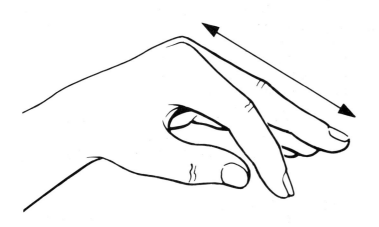

of standard material, gives a test of a useful ability; dry seeds—peas, sunflower 'seeds', maize, or (more accurate but more demanding) dry sand—can be used. The longest hands may well not be the 'winners'. How big *is* a handful?

Human 'wing-span', from finger-tip to finger-tip against the wall, is a measurement of special interest when compared with the height of the owner. Of course, some people have long arms for their 'size', while others have unusually short ones; both groups are likely to know about their difficulties with sleeves. However, the general pattern holds, and is useful when, for example, investigating the rough sizes of rooms in a possible new house or flat.

Upward reach is worth measuring, best done against a vertical tape-measure type scale on the wall. It can be one-handed, useful for bolting the back door, or two-handed, for getting things off high shelves. . . upward leap is probably measured most easily from the marks made by (damp) fingertips on a vertical chalkboard.

Actual hand length is not easy because of the wrists attached; some agreement must be worked out as to where the hand ends and the wrist begins. This is equally true for hand volume measurement, i.e. 'How far in?' when displacing water, and for the area of a hand on squared paper. This is a difficult project in any case because of likely inaccuracies in drawing the outline and problems in counting the part-squares.

Distances round

Distances round—necks, waists, etc.—are closely connected with clothing sizes, and are interesting. They need to be approached with sensitivity, since some children—and adolescents—are very conscious of being above, or below, the mean, or mode, of their colleagues. However, a class can pool information and collect clothing size labels which link English measurements and sizes with Continental and perhaps also American systems (and vocabulary).

The most tactful approach to 'chest measurement', apart from clothing sizes, is probably to ignore the simple circumference, and to concentrate on the difference between the measurement with a large breath taken in, and with as much air breathed out as possible—i.e. chest expansion. This may well have no direct relationship to circumference, but may perhaps be found to have some connection with training in activities such as swimming. Like so many other measurements of human beings, this is one where increased effort may change the result, and even the second attempt may be measurably different from the first. When converting the results for a class into a graphical form, it is often most convenient to group the measurements, e.g. less than 1 cm; between 1 and 2 cm; between 2 and 3 cm and so on. This is, of course, the practice employed by clothing manufacturers too.

Sitting height

Other length measurements may be suggested, and some will have everyday importance, such as sitting height. This really matters for many people, among them typists and car-drivers. People with short legs may have long backs, so standing height is not a good test. The height of the seat of a hard upright chair can be marked on the wall, and that of the top of the individual's head found as for standing height.

Reading distance

Similarly important, both for children and for adults using fine print, microfiches or VDUs, is comfortable reading distance. This can only be found by experiment, say with the phone book; older people sometimes say in fun that it is not their sight which makes this difficult, but that their arms are not long enough any more.

Hair

Children are often fascinated by the length to which uncut human hair will grow; it used sometimes to be said—in surprise and praise—'Her hair is so long that she can sit on it.' Maybe someone in the school can achieve this distinction—and can be persuaded to spare just one hair for scientific measurement?

Fingernails

As for the growth of fingernails (and toenails?—see the play *Rosencrantz and Guildenstern are dead*, in which the two characters discuss it at length) a small scratched mark shows the current position of the point where nail and finger separate, and measurement (in millimetres) after one week gives the extension. If one of the group knows someone—probably a young woman—with very long nails, or perhaps just one little-finger nail, they could ask her tactfully if they might measure it.

Weights

What about weights? Here again a little tact may be needed, but each individual can find his or her own weight without announcing it. The ordinary bathroom scales are appropriate, but they must be on a flat floor and need adjusting to zero, preferably after each use. The built-in magnifier is a point of interest; why does this particular instrument have one? Maybe so that we don't have to crouch to read the scale? Attention can be distracted from the under-weights of girls who tend towards anorexia, for example, by finding as an exercise the weights of different kinds of cold-weather outer clothing, or of wellies and football boots. After all, both of these have to be lifted with every step, and it would not be surprising to find them producing tired leg muscles. Since the bathroom-type scales are not extremely accurate, adding the smaller weight, e.g. the boots, to the person already on the scale platform usually gives a better result than weighing the small object alone.

One could try the weights of individuals with and without other things which are carried as a matter of course, such as the bag or briefcase—which often contains unnecessary books, or even a load of conkers? or marbles? or the fruit uneaten at lunchtime. . . . In some classes it is possible without fuss to let pupils find their own weight before and after going to the toilet—after all, it must make a difference, mustn't it?

Pupils may know that jockeys and boxers have to get their weights into very strict limits—maybe by sweating and dieting. Boxers in the featherweight class, for example, are weighed in their shorts at 11 am for an afternoon match, and must be below 57.153 kg, which means that their weight matters down to one gram!

Babies are usually weighed regularly, and therefore provide good information for studying growth. The very youngest may, of course, lose slightly in their first week or so, but after that they are checked to see how soon they double their birth-weight, and this may give a lively incentive to their older siblings. These may also be able to supply their own 'statistics', and to see how many times they have increased their original weight in the years between.

If no family baby seems available for study, there may be a teacher on maternity leave who would help, and she could maintain her link with her class in a way which would give pleasure all round if by arrangement she brought the baby for measuring.

Timing

Pulse rate

Many of the activities commonly suggested for timing, such as pulse rate (which is, of course, the same thing as heart rate), are very variable. This means that when we try to measure them, we need to do several checks, and to try to identify reasons for discovered differences. The actual position of the wrist pulse is sometimes difficult to find; it may be easier to feel just above the collar-bone, or to watch for the movement of the leg crossed over its twin just above the knee. A shiny ear-clip (not a ring through a pierced ear) will sometimes give a very clear double movement (the 'lubb-dup' of medical descriptions of heart-beat) which in sunlight can show a wide reflected movement on the wall. Timing has to depend on the movement being felt or seen; some testers press too lightly at first, and some so heavily that they stop the blood-flow in that particular blood-vessel altogether in trying to find it. A clock with a third 'sweep' hand gives each individual a chance to begin timing when the pulse has been located.

Breathing rate

Breathing rates are interesting to time, but are very controllable by the owner; for example, a few deep breaths can provide as much ventilation as many very shallow ones. Previous activity is well known to affect the breathing rate, but it may go in either direction—faster and perhaps shallower, or slower and deeper. This is one of the many characteristics of living things which demonstrates the differences between individuals—and is valuable in this respect. The same goes for the length of time one can hold one's breath—though this should never be taken to extremes.

Blinking

Blinking is another easily measured activity, timed by the person or by an observer. Some very young children—perhaps the babies of the previous section—seem able to stare without blinking for almost embarrassingly long periods, and without any inducement.

Angles in people

Angles are quite relevant to human anatomy and

movements, though they can seldom be measured very accurately, and again are very variable from person to person. Some, however, are interesting from other points of view than simple measurement. Consider some of the following:

(a) complete turning of the head, without moving the shoulders;

(b) angle of vision from side to side without moving the head;

(c) mobility of arm at the shoulder joint as in overarm bowling (N.B. girls can do this too, in spite of some prejudice);

(d) elbow joint movement—may go a little past the 180° position;

(e) finger joints—mobility to or beyond the straight line varies very much with the individual (though 'double-jointedness' does not exist);

(f) the special quality of the joint at the base of the thumb—it may be possible to 'draw circles' with it, as with ankles;

(g) knee joints—which rarely go past the 180° straight line.

Some pupils who have ballet training can do much more with their joints than those who have had no stretching exercises—some may even have learnt and stretched enough to do 'the splits', but of course this kind of extreme position can be dangerous if not approached through long and skilful training, and emulation by the untrained has to be prevented.
Measurement methods can be devised, often using two rulers to make the angle to match the limb, and the resulting angle being measured with a large protractor, or drawn on paper and a normal protractor used.

Drawing attention to the angles which limbs can make gives a far greater understanding of the anatomy of common animals. For example, a horse cannot really have knees in its front legs and because elbows don't bend that way the joints sometimes protected with leather pads must be 'backs of wrists'. The horse's or dog's or cat's knees are tucked up beside the animal's body, almost unseen, while the backward-extending joint has to be the ankle or heel. Measuring is possible with a co-operative pet.

Measuring plants

In the classroom

There are many more or less mechanical measurements suggested in books with reference to plants growing in classrooms; the number of germination successes, dates of germination from sowing date, height records for seedlings, heights of climbing plants, areas of leaves and their growth, comparative studies of roots, and so on.

These may become, even if they were not so intended, competitive events. Competitive events can be positively valuable, given as fair a start as possible, e.g. similar seeds and similar external conditions (not, as in one book, some children being given baked beans to plant. . .) and where success depends mainly on care and common sense.

Out of doors

The adult Horticultural Show or Gardening Society Exhibition—right up to Chelsea—depends quite largely on comparative measurements. Plants grown at school are often over-watered out of enthusiasm, but if the site is chosen carefully and provided with good drainage, and the teacher is forearmed with slug poison, gratifying results may be measured. Marrows can be used as examples of growth:time relationships, and can even be weighed on the plant if they are growing on a bank or in a raised bed.

Other choices could be the tallest hollyhock! the heaviest potato crop from one 'seed' potato; the longest flowering season for a chosen species (weeds are good for this—that's one of the ways they get to be weeds. . .); and the highest sunflower head. Some of these call for ingenuity in devising methods of measurement, e.g. the 3-metre hollyhock during its growth.

A row of sunflower seeds along a school fence, away from vandals and footballs, will offer much material; such as the number of leaves each week during early growth, heights, angles between stem and leafstalk (the sunflower is adapted to make the most of sunlight), the rotation of the flower-head with the direction of sunlight (does it turn on dull days? does it start in the morning where it left off last night?), the number of bees visiting the flowerhead in (say) 10 minutes, the width of the fully-grown head, the height of the stem at its tallest. When the fruits are ripe and dry, what weight was produced, and how many fertilized fruits ('sunflower seeds') were there altogether? The easiest way to get this figure is to divide the crop roughly between a number of pairs of pupils, and to use a calculator to add up the totals—with a reminder to the census-takers to squeeze each fruit gently between finger and thumb to make sure that there really is a seed inside the tough skin.

Many other easily-grown plants, such as peas or runner beans, can offer similar possibilities, with a harvest of some sort at the end. . . Local fruiting trees—from horse-chestnuts and oaks to (cooking?) apples—offer other chances to make meaningful measurements. It is striking to discover what weight of 'protection' a tree such as a horse-chestnut actually provides in the way of a

prickly case for its precious contents when they hit the ground. . . The weight of a cultivated apple compared with the weight of the pips inside is due to human selection, but the 'conker case' is certainly the result of natural causes. Fruit is usually sold by weight; what weight of orange peel or banana skin do we buy, pay for, and throw away?

On the other hand, if a local tree is being cut down, one can marvel at the weight of wood produced from a seed of which the approximate weight can be found in the autumn.

Measuring live animals

There may be built-in disadvantages to measuring animals—the creature does not co-operate, is very small, very quick to move or perhaps is very large.

Minibeasts

Minibeasts such as wood-lice are often suggested as raw material for many kinds of investigation, so what can we measure about these tiny crustaceans (not insects, of course)? We can measure length using centimetre squared paper or a plastic grid; this is much easier than expecting them to stand still on a ruler, facing the right way. One can get a rough idea of the length of an ant or a ladybird on a centimetre grid—or even better, on the 5 millimetre squared paper which is much easier to find in stationers' shops. Measuring the length of earthworms, even if described by Nuffield texts, is not recommended. The hands of competitive small boys are *not* good for earthworms; in any case, the worms tend to dry, and hence to die of suffocation.

A problem arises: do lengths of minibeasts include legs, e.g. spiders, crane flies (Daddy-long-legs); or antennae, e.g. shrimps? Measurement from where to where needs to be decided in advance.

Average (mean) weights can be found if there are plenty of relatively slow-moving specimens and a sensitive balance. If a small quantity of 'habitat' is thought to be a good thing, to 'keep them happy', this can be weighed in the balance 'tub' first.

Rate of locomotion

Speed, or rate of locomotion, is fraught with difficulties. 'How fast does a snail travel' is really unanswerable, and children's ideas of 'encouragement' with pencil-points, etc. are not acceptable. What can be done is to allow active minibeasts to wander between walls, e.g. of Plasticine on squared paper (or acetate grid) and observe the distance covered in a definite length of time. The same arrangement can be used to find the lengths of earthworms if this is wanted, with a drop or two of water in the 'gully'.

Family pets

These usually offer good material for measurements, and may enjoy being the centre of attention (and petting). Two or three dogs, taken one at a session, can give a very instructive range of information:

(a) shoulder height (as for horses, but in decimal units . . .) comparing dachshund, whippet, spaniel, mongrel of any background, etc;

(b) ear length (not countenancing ear-cropping);

(c) weight (either in bucket-type scales or in owner's arms on the bathroom scales);

(d) lengths of coat-hairs (from combings), showing the two types—long guard hairs and short under-coat providing warmth—in some breeds;

(e) total length—head, body, tail—in dogs which still have their full tail-length;

(f) and comparative studies of several of these measurements in pairs, e.g. length/height, head/body, height/weight.

Growth rates, especially weights, are very interesting, and if a family of young animals, say a litter of puppies, is available, scientific data can be collected and graphed to the satisfaction of all concerned.

And others . . .

Schools in the country, or even with a nearby friendly riding-school, have many more possibilities than inner-city schools.

The relative heights and lengths of ponies and horses of different breeds are often well-known already. Worn-out shoes give an accurate indication of foot-size—how far does this correlate with height? Farmers who still have working horses may know their weights too, and the growth rate of the new foal is of real interest. The weights of piglets are often checked regularly by the farm staff—up to selling weight. . . Dairy farming involves a good deal of measurement and record-keeping, which with tact and intelligent pupil-behaviour can often be observed.

Measurement should throughout be seen as an activity with meaning, as in many of the examples in this book. It is not intended to be a mechanical exercise, but as something of general importance in the real world, and therefore to be done with accuracy and understanding.

Index

72